Wissenschaftliche Reihe Fahrzeugtechnik Universität Stuttgart

Reihe herausgegeben von

André Casal Kulzer, Stuttgart, Deutschland

Hans-Christian Reuss, Stuttgart, Deutschland

Andreas Wagner, Stuttgart, Deutschland

Das Institut für Fahrzeugtechnik Stuttgart (IFS) an der Universität Stuttgart forscht interdisziplinär sowie technologieoffen an modernen und zukunftsorientierten Fahrzeugkonzepten. In enger Zusammenarbeit mit Partnern aus Industrie und Wissenschaft entstehen neue Lösungen für die Mobilität der Zukunft. Das Institut gliedert sich in drei spezialisierte Lehrstühle, die gemeinsam das gesamte Spektrum der Fahrzeugtechnik abdecken:Der **Lehrstuhl für Fahrzeugantriebssysteme** widmet sich der Forschung nachhaltiger Antriebslösungen für künftige Mobilitätskonzepte. Im Fokus stehen alternative, elektrische sowie hybride Antriebssysteme und deren Komponenten – einschließlich der Nutzung nachhaltiger Energieträger wie Wasserstoff, synthetischer Kraftstoffe und Batterien. Der **Lehrstuhl für Kraftfahrzeugmechatronik** beschäftigt sich mit vernetzten, intelligenten und adaptiven Fahrzeugen. Im Zentrum stehen Fragestellungen zum Automatisierten und Vernetzten Fahren, Diagnose, Ladetechnologien, verteilte Systeme sowie softwarebasierte Fahrzeugfunktionen. Der **Lehrstuhl für Kraftfahrwesen** erforscht die physikalischen Grundlagen der Auslegung zukünftiger Fahrzeugkonzepte. Im Mittelpunkt stehen die Bereiche Aerodynamik, Windkanaltechnik, Akustik/NVH, Fahrzeugdynamik, Reifenmanagement und Thermomanagement Gesamtfahrzeug. Das IFS verfügt über eine vielfältige und hochmoderne Forschungsinfrastruktur, die realitätsnahe Untersuchungen vom Einzelbauteil bis zum Gesamtfahrzeug ermöglicht. Besonders hervorzuheben sind der Multikonfigurations- und Antriebsprüfstand für komplexe Antriebskonzepte, der Stuttgarter Fahrsimulator zur Untersuchung menschlichen Fahrverhaltens, der Aeroakustik-Fahrzeugwindkanal für akustische und strömungstechnische Fragestellungen sowie der Thermowindkanal zur Analyse thermischer Prozesse im Gesamtfahrzeug. Die wissenschaftliche Reihe „Fahrzeugtechnik Universität Stuttgart" dokumentiert die im Rahmen von Promotionen am IFS entstandene Beiträge zur Mobilität der Zukunft und zeigt deren thematische Vielfalt, methodische Tiefe und Praxisrelevanz.

Reihe herausgegeben von

Prof. Dr.-Ing. André Casal Kulzer
Lehrstuhl für Fahrzeugantriebssysteme
Institut für Fahrzeugtechnik Stuttgart
Universität Stuttgart
Stuttgart, Deutschland

Prof. Dr.-Ing. Andreas Wagner
Lehrstuhl für Kraftfahrwesen
Institut für Fahrzeugtechnik Stuttgart
Universität Stuttgart
Stuttgart, Deutschland

Prof. Dr.-Ing. Hans-Christian Reuss
Lehrstuhl für
Kraftfahrzeugmechatronik
Institut für Fahrzeugtechnik Stuttgart
Universität Stuttgart
Stuttgart, Deutschland

Carlo Alberto Perugini

An Efficient Hybrid Computational Process for Aeroacoustic Vehicle Development Based on Navier-Stokes Equations and Lighthill's Acoustic Analogy

Springer Vieweg

Carlo Alberto Perugini
Institute of Automotive Engineering
(IFS), Chair of Automotive Engineering
University of Stuttgart
Stuttgart, Germany

Zugl.: Dissertation Universität Stuttgart, 2026
D93

ISSN 2567-0042   ISSN 2567-0352   (electronic)
Wissenschaftliche Reihe Fahrzeugtechnik Universität Stuttgart
ISBN 978-3-658-51976-6   ISBN 978-3-658-51977-3   (eBook)
https://doi.org/10.1007/978-3-658-51977-3

This Springer Vieweg imprint is published by the registered company Springer Fachmedien
Wiesbaden GmbH, part of Springer Nature.
The registered company address is: Abraham-Lincoln-Str. 46, 65189 Wiesbaden, Germany

If disposing of this product, please recycle the paper.

Acknowledgements

I would like to express my sincere gratitude to my supervisor, Prof. Dr.-Ing. Andreas Wagner, for his guidance and mentorship. His critical perspective and insightful discussions have greatly contributed to my personal and professional growth.

I would also like to thank Prof. Simone Sebben for taking on the role of co-referee and the effort involved, as well as for her valuable comments.

My heartfelt thanks go to Reinhard Blumrich and Domenic Staron for their endless professional and personal support. I am equally grateful to all my colleagues at FKFS and Lamborghini, whose collaboration has been crucial throughout the entire research journey.

I would also like to extend a special thank you to my Stuttgart family - Antonio, Antonino, Cristian, Edoardo, Francesco, Greta, and Mario - for their presence and support during this time.

A warm thank you also goes to my lifelong friends - too many to name, but you know who you are - your constant presence has always been a steady reference point in my life.

I am also grateful to Francesco and Daniele, whose example - though distant - has guided me over the years and has been a lasting source of inspiration in the pursuit of my goals.

To my mother, my father, and my aunt, who have been by my side from the very beginning, I owe my deepest gratitude for their unwavering love and for being my greatest supporter.

Last but not least, I wish to thank my partner in laughter Francesca for sharing both the joys and the challenges of this journey, for the countless nights spent working together, but above all, for always sharing her smile with me.

Carlo Alberto Perugini

Table of Contents

Figures & Tables

Symbols

B	-	Empirical constant
c	m/s	Speed of sound
c_{b1}	-	Model constant in Spalart-Allmaras model
c_{b2}	-	Model constant in Spalart-Allmaras model
C_{DES}	-	Empirical constant in DES model
$c_{P,stat}$	-	Static pressure coefficient
c_{w1}	-	Model constant in Spalart-Allmaras model
d	m	Distance to the nearest wall
d_{DES}	m	Modified length scale in DES model
$\tilde{d}_{DES}$	m	Modified length scale in DDES model
E	Pa	Young modulus
f	Hz	Frequency
f_d	Hz	Shielding function in DDES model
f_{max}	Hz	Maximum frequency
f_s	Hz	Sampling frequency
f_w	-	Damping function
h_s	m	Glass thickness
I_d	W/m^2	Sound intensity for dipole sources
I_m	W/m^2	Sound intensity for monopole sources
I_q	W/m^2	Sound intensity for quadrupole sources
k	m^2/s^2	Turbulent kinetic energy
k_a	1/m	Wavenumber related to the acoustic fluctuation

k_b	1/m	Structural bending wavelength of the window
k_h	1/m	Wavenumber related to the hydrodynamic fluctuation
l	m	Hydrodynamic turbulence scales
L_{max}	m	Maximum size of the FE mesh
Ma	-	Mach number
N	-	Number of samples used in the DFT
$\hat{n}$	-	Normal vector to the solid surface
p	Pa	Pressure
$\bar{p}$		Time independent pressure component
p'		Time fluctuating pressure component
p_{dyn}	Pa	Dynamic pressure
p_{rms}	Pa	Root mean square sound pressure
p_{stat}	Pa	Static pressure
p_{tot}	Pa	Total pressure
p_0	Pa	Reference pressure, typically $2 \cdot 10^{-5}$ Pa
Q	s^{-2}	Second invariant of the velocity gradient tensor
r_d	-	Dimensionless parameter in DDES shielding function
Re	-	Reynolds number
RT_{60}	s	Reverberation time
S	s^{-1}	Vorticity magnitude
$\tilde{S}$	s^{-1}	Intermediate variable, function of vorticity magnitude and modified turbulence viscosity

t	s	time
T_{DFT}	s	DFT window length
T_{ij}	Pa	Lighthill's stress tensor
U	m/s	Flow velocity
U_{mag}	m/s	Velocity magnitude
U_{max}	m/s	Maximum expected velocity
U_∞	m/s	Free stream velocity
U_τ	m/s	Friction velocity
u, v, w	m/s	Components of the flow velocity in the x, y, z directions, respectively
$\bar{u}, \bar{v}, \bar{w}$	m/s	Time independent velocity components
u', v', w'	m/s	Time fluctuating velocity components
$\tilde{u}, \tilde{v}, \tilde{w}$	m/s	Resolved scales of velocity components
$u_{SGS}, v_{SGS}, w_{SGS}$	m/s	Subgrid scales of velocity components
u^+, y^+		Dimensionless velocity and wall distance
x, y, z	m	Cartesian coordinates
δ	m	Thickness of the boundary layer
δ_{ij}	-	Kronecker delta
$\delta\rho$	Kg/m^3	Test function in Lighthill's variational formulation
Δ	m	Largest CFD grid spacing in all three directions
Δ_{min}	m	Smallest CFD grid spacing in all three directions
Δt	s	CFD timestep
Δt_{ext}	s	CFD velocity field extraction timestep
$\Delta x, \Delta y, \Delta z$	m	CFD grid spacing components

Symbol	Unit	Description
ε	m²/s³	Dissipation rate of turbulent kinetic energy
ε_s	-	Poisson ratio
$\varepsilon_\%$	-	Percentage error
θ	rad	Directional dependence on pressure radiation
κ	-	Von Kármán constant, approximately 0.41
λ	m	Acoustic wavelength
λ_{min}	m	Smallest acoustic wavelength
μ	Pa·s	Dynamic viscosity
ν	m²/s	Kinematic viscosity
$\tilde{\nu}$	m²/s	Modified turbulence viscosity
ρ	Kg/m³	Fluid density
ρ_s	Kg/m³	Solid density
ρ_0	Kg/m³	Reference fluid density
τ_{ij}	Pa	Viscous stress tensor
τ'_{ij}	Pa	Reynolds stress tensor
ϕ	Pa	Filtered source term
$\phi_c(f)$	-	Frequency dependent cabin sound absorption coefficient
Ω	m³	Arbitrary integration volume in Finite Element

Abbreviations

AA	Acoustic Analogies
APE	Acoustic Perturbation Equations
AVA	Aero-Vibro-Acoustic
BCs	Boundary Conditions
BEM	Boundary Elements Method
CAA	Computational Aeroacoustics
CFD	Computational Fluid Dynamics
CFL	Courant-Friedrichs-Lewy
CPU	Central Processing Unit
CPUh	Central Processing Unit per Hours
DDES	Delayed Detached Eddies Simulations
DFT	Discrete Fourier Transform
DES	Detached Eddies Simulations
DNC	Direct Noise Computation
DNS	Direct Numerical Simulation
H-R	High Reynolds
HSM	Hyundai Simplified Model
IDDES	Improved Delayed Detached Eddies Simulations
FE(M)	Finite Elements (Method)
FKFS	Forschungsinstitut für Kraftfahrwesen und Fahrzeugmotoren Stuttgart
FVM	Finite Volume Method
FW-H	Ffowcs Williams and Hawkings
LBM	Lattice-Boltzmann Method
LEE	Linearized Euler Equations

LES	Large Eddies Simulation
LHS	Left-Hand Side
L-R	Low-Reynolds
NSE	Navier-Stokes Equations
NVH	Noise Vibration and Harshness
OEM	Original Equipment Manufactures
OSPL	Overall Sound Pressure Level
PCE	Perturbed Compressible Equations
PE	Perturbation Equations
PFL	Pressure Fluctuation Level
RANS	Reynolds-Averaged Navier-Stokes
RHS	Right-Hand Side
RSM	Reynolds Stress Models
S-A	Spalart-Allmaras
SAE	Society of Automotive Engineers
SEA	Statistical Energy Analysis
SGS	Subgrid Scale
SPL	Sound Pressure Level
SUV	Sport Utility Vehicle
TFI	Turbulent Flow Instrumentation
TKE	Turbulent Kinetic Energy
URANS	Unsteady Reynolds-Averaged Navier-Stokes
VA(M)	Vibro-Acoustic (Model)
WT	Wind Tunnel

Zusammenfassung

Die Automobilindustrie sieht sich stetig wachsenden Anforderungen gegenüber, unter anderem im Hinblick auf die Sicherheit der Fahrzeuginsassen und in Bezug auf deren akustischen Komfort. Neben regulatorischen Vorgaben und steigenden Kundenerwartungen spielt dabei insbesondere die subjektive Wahrnehmung von Geräuschen eine zunehmend zentrale Rolle in der Fahrzeugentwicklung. Die Reduktion von störenden Geräuschen im Fahrzeuginneren ist dabei ein zentrales Anliegen, insbesondere vor dem Hintergrund neuer Fahrzeugkonzepte. Mit zunehmender Geschwindigkeit - typischerweise über 100 km/h - wird das aerodynamische Geräusch zu einer der dominierenden Lärmquellen. Dieses entsteht durch die komplexen und instationären Turbulenzen, die sich im Strömungsfeld rund um das Fahrzeug ausbilden. Diese Strömungsphänomene sind stark nichtlinear und stellen sowohl experimentell als auch numerisch eine erhebliche Herausforderung dar.

Diese Problematik gewinnt besonders bei Hybrid- und Elektrofahrzeugen an Bedeutung. Da bei diesen Antriebskonzepten das Geräusch des klassischen Verbrennungsmotors zumindest teilweise entfällt und der Elektromotor deutlich leiser ist, verschiebt sich die akustische Wahrnehmung der Insassen zunehmend hin zu aerodynamisch verursachten Geräuschanteilen (wobei auch Reifen-Fahrbahn-Geräusche an Bedeutung gewinnen). Entsprechend steigt der Bedarf an gezielter aeroakustischer Optimierung in der Fahrzeugentwicklung. Dies führt zu einer verstärkten Integration akustischer Kriterien bereits in frühen Entwicklungsphasen.

In diesem Zusammenhang ist die Computational Aeroacoustics (CAA) auf dem Weg, sich als ein unverzichtbares Instrument zu etablieren. Sie ermöglicht detaillierte Einblicke in das aeroakustische Verhalten von Fahrzeugkomponenten - und dies bereits in frühen Phasen der Produktentwicklung auf Basis virtueller Prototypen. Dank dieser Methodik können Entwicklungszyklen verkürzt, reale Prototypen eingespart und Kosten reduziert werden. Dennoch bringt die Anwendung von CAA in der Praxis einige Herausforderungen mit sich: Vor allem die hohen Rechenkosten und der Bedarf an Speicherkapazität begrenzen ihren durchgängigen Einsatz im industriellen Umfeld. Eine zentrale Aufgabe besteht daher darin, einen tragfähigen Kompromiss zwischen physikalischer Genauigkeit und numerischer Effizienz zu finden, um die Integration der Methode in reale Entwicklungsprozesse zu ermöglichen. Insbesondere im industriellen Kontext ist eine robuste und reproduzierbare Methodik von entscheidender Bedeutung.

Hybride numerische Verfahren haben sich in diesem Spannungsfeld als besonders geeignet erwiesen. Sie kombinieren unterschiedliche Modellierungsmethoden, um den jeweiligen Stärken gerecht zu werden und gleichzeitig deren Schwächen zu kompensieren. In der wissenschaftlichen Literatur finden sich zahlreiche Studien, die sich mit solchen hybriden Verfahren befassen. Jedoch beziehen sich viele dieser Arbeiten auf idealisierte oder stark vereinfachte, generische Fahrzeuggeometrien. Damit fehlt es an ausführlich dokumentierten Anwendungsfällen, welche die Modellierung und Validierung für reale Serienfahrzeuge - mit all ihrer geometrischen und physikalischen Komplexität - systematisch darstellen. Ebenso unzureichend ist bislang die Beschreibung standardisierter Workflows, die eine Einbindung in die industrielle Fahrzeugentwicklung erlauben würden. Auch die Kombination numerischer Modelle mit experimentellen Windkanal–Messungen in Bezug auf Fahrzeugoptimierung wurde bisher kaum umfassend dokumentiert. Gerade diese Lücke stellt jedoch einen entscheidenden Faktor für die industrielle Anwendbarkeit dar.

Ein weiterer kritischer Punkt ist die Fokussierung der meisten bisherigen Studien auf den Außenspiegel als aeroakustische Hauptquelle. Obwohl der Außenspiegel unbestritten ein relevanter Bereich für Windgeräusche ist, bleiben andere, potenziell ebenso kritische Regionen weitgehend unbeachtet. Besonders im hinteren Fahrzeugbereich, etwa bei Bauteilen wie Dachspoiler, Antennen oder der Heckscheibe, treten oft komplexe Effekte durch aerodynamische Wechselwirkungen auf. Diese führen zu nichtlinearen akustischen Phänomenen, deren Modellierung und Analyse mit erheblichen Herausforderungen verbunden ist. In der Standardkonfiguration vieler Fahrzeuge gelten diese Bereiche zwar als weniger kritisch - unter spezifischen Umständen, wie etwa bei Fahrzeugen mit sportlichen Heckanbauten, ausgeprägten Kanten oder sehr hohen Geschwindigkeiten, kann die Relevanz ihres aerodynamischen Geräusches jedoch stark zunehmen. Dennoch existieren bislang kaum Studien, die sich systematisch mit diesen Einflussbereichen auseinandersetzen. Eine ganzheitliche Betrachtung des gesamten Fahrzeugs ist daher erforderlich, um alle relevanten Geräuschquellen zu erfassen.

Im Rahmen dieser Arbeit wurde ein dreistufiges hybrides Simulationsverfahren entwickelt, das folgende Komponenten integriert: eine numerische Strömungssimulation mittels Computational Fluid Dynamics (CFD), ein Finite-Elemente-Modell (FEM) zur Vorhersage der akustischen Schallabstrahlung sowie ein vibroakustisches Modell (VAM) zur Simulation der Geräuschübertragung in den Fahrzeuginnenraum. Jeder dieser Schritte baut auf den Ergebnissen des vorangegangenen auf, sodass ein durchgängiger, konsistenter Simulationsprozess entsteht. Dieser modulare Aufbau ermöglicht zudem eine flexible Anpassung an unterschiedliche Anwendungsfälle.

Die CFD-Simulation wurde in dem Navier-Stokes-Solver OpenFOAM durchgeführt, unter der Annahme inkompressibler Strömung und unter Verwendung des DDES-Ansatzes (Delayed Detached-Eddy Simulation), basierend auf dem Spalart-Allmaras Turbulenzmodell. Dieses Modell ermöglicht eine zuverlässige Erfassung großskaliger, instationärer Strukturen im Strömungsfeld - bei deutlich geringerem Rechenaufwand als bei einer hochaufgelösten Large-Eddy-Simulation (LES). Die berechneten Geschwindigkeitsfelder dienen anschließend als Eingabedaten für die FEM-Simulation. Diese basiert auf Lighthills akustischer Analogie sowie auf den Acoustic Perturbation Equations (APE), um den Außenschall effizient bestimmen zu können. Im letzten Schritt wird die Übertragung des aerodynamischen Geräuschs durch die Fahrzeugscheiben mittels VAM simuliert. Hierbei werden die durch das FEM berechneten Druckfluktuationen als Randbedingung verwendet. Die FEM- und VAM-Berechnungen wurden mithilfe der Software Actran realisiert, wodurch die akustischen Effekte abgebildet werden konnten. Durch diese Kopplung wird eine durchgängige physikalische Beschreibung des gesamten Übertragungswegs erreicht.

Untersuchungsobjekt war ein Lamborghini Urus SUV bei einer konstanten Geschwindigkeit von 140 km/h und einem Anströmwinkel von 0°. Die Analyse konzentrierte sich auf zwei Schlüsselbereiche: den Außenspiegel sowie den Dachspoiler. Der untersuchte Frequenzbereich umfasste 100 Hz bis 5000 Hz, womit der größte Teil der aeroakustisch relevanten Frequenzen berücksichtigt wurde. Die Simulationsergebnisse wurden anhand experimenteller Daten validiert, die im aeroakustischen Windkanal der Universität Stuttgart gewonnen wurden. Um reproduzierbare Ergebnisse zu gewährleisten, wurden zwei gleiche Fahrzeuge unter exakt gleichen Bedingungen vermessen. Dabei wurde großer Wert darauf gelegt, sowohl außen als auch innen eine realitätsnahe, aber auch simulationskonforme Messumgebung zu schaffen. Zur

Vermeidung externer Störeinflüsse und zur gezielten Untersuchung der durch die Geometrie eines definierten Fahrzeugbereichs verursachten Geräuschquellen wurden alle Dichtungen abgeklebt und eine spezielle Innenraumdämmung installiert.

Der Validierungsprozess war dreistufig aufgebaut - analog zur Simulationskette. Zunächst wurden die CFD-Ergebnisse hinsichtlich Geschwindigkeits- und Druckverteilungen mit den Zeitmittelwerten aus den Messdaten verglichen. Anschließend erfolgte ein Abgleich der Schalldruckpegel im Außenraum durch Mikrofone, die gezielt außerhalb der direkten Turbulenzzonen des Nachlaufs von Spiegel und Spoiler positioniert wurden. Abschließend wurden Innenraummikrofone an Fahrer- und Rücksitzplatz verwendet, um die Schalleinträge durch die Seitenscheiben mit den VAM-Vorhersagen zu vergleichen. Der gesamte Validierungsprozess wurde umfassend dokumentiert, ist nachvollziehbar und für andere numerische Verfahren adaptierbar.

Die Ergebnisse zeigten eine sehr gute Übereinstimmung zwischen Simulation und Messung im gesamten Frequenzbereich. Lediglich bei tiefen Frequenzen kam es zu kleineren Abweichungen. Diese waren im Außenbereich auf die begrenzte Auflösung des Simulationsmodells im niederfrequenten Bereich zurückzuführen. Im Innenraum wurden Unterschiede durch vereinfachte Modellannahmen bei der Schallübertragung - insbesondere im Hinblick auf Fensterabdichtungen und die Beschreibung der Einspannung - verursacht. In der Simulation wurden die Scheiben als frei schwingend angenommen, wodurch insbesondere bei tiefen Frequenzen eine Überschätzung der Vibrationen entstand.

Weiterführende Analysen zeigten, dass das Modell im tieffrequenten Bereich eine hohe Sensitivität gegenüber der physikalischen Zeitdauer der Simulation sowie der Art der verwendeten Randbedingungen aufwies - insbesondere im Hinblick auf die Anregung der Seitenscheibe. Besonders deutlich wurde dies bei der Frage, ob hydrodynamische Druckschwankungen als Eingang in das VAM einbezogen werden. Ohne deren Berücksichtigung - also bei Verwendung ausschließlich akustischer Fluktuationen - wurde das tieffrequente Innenraumgeräusch deutlich unterschätzt. Eine Wellzahlanalyse bestätigte diese Beobachtung: Unterhalb von 500 Hz dominieren hydrodynamische Beiträge, während im Bereich zwischen 500 Hz und 5 kHz die akustischen Anteile überwiegen.

Das vorgestellte Verfahren zeichnet sich durch hohe Genauigkeit bei vergleichsweise geringem Rechenaufwand aus. Es eignet sich für die gezielte Untersuchung unterschiedlicher Fahrzeugbereiche mit je eigenen Strömungsbedingungen. Jeder Schritt der Simulation trägt spezifisch zur Gesamterkenntnis bei: Die Strömungsanalyse liefert ein Verständnis der Geräuschentstehung, die akustische Analyse im Außenfeld erlaubt die Lokalisierung von Schallquellen und das vibroakustische Modell ermöglicht eine Bewertung der Innenraumakustik aus Sicht der Insassen.

Die Effizienz des Verfahrens ergibt sich unter anderem aus der Annahme inkompressibler Strömung. Zwar schränkt diese Annahme die Möglichkeit ein, bestimmte aerodynamisch-akustische Kopplungseffekte zu modellieren, sie reduziert jedoch den Gesamtaufwand erheblich. Die CFD bleibt dabei der aufwendigste Simulationsschritt - sowohl hinsichtlich Rechenzeit als auch hinsichtlich Speicherbedarf. Zur praktischen Anwendung ist deshalb eine weitgehende Automatisierung notwendig, um den Zeitbedarf weiter zu senken. Der manuelle Aufbau der CFD-, FEM- und VAM-Modelle für jede Konfiguration führt zu einem erheblichen Mehraufwand.

Eine abschließende Analyse der aeroakustischen Quellen diente einer vertieften Untersuchung der Entstehungsmechanismen und dem besseren Verständnis der Natur aerodynamischer Geräusche. Dazu wurde eine Kombination aus numerischer Vorhersage und experimenteller Array-Messung verwendet. Letztere ermöglichte eine gezielte räumliche Zuordnung der Schallquellen, während die CFD-Analyse eine detaillierte Charakterisierung der Wirbelstrukturen lieferte, die maßgeblich zur Geräuscherzeugung beitragen. Eine hohe Übereinstimmung hinsichtlich der Schalldruckverteilung im Bereich des Außenspiegels und Dachspoilers konnte nachgewiesen werden.

Beim Außenspiegel zeigte sich, dass die Hauptquelle des Außengeräuschs auf die Wirbelablösung im Bereich des Spiegeldreiecks zurückzuführen ist – eine Region mit starker Wechselwirkung zwischen Strömung und Struktur. Gleichzeitig wurde dieser Bereich auch als Hauptquelle des Innengeräuschs identifiziert, was sowohl durch Array-Messungen und Innenraummikrofone als auch durch VAM-Prognosen bestätigt wurde.

In der Analyse des Dachspoilers wurden zwei Konfigurationen untersucht: ein Standardspoiler sowie eine sportliche Variante. Beide Varianten erzeugten hochgradig turbulente Strömungsfelder auf der Heckscheibe, die dominanten Schallquellen im Bereich der Rücksitzplätze darstellten. Diese beeinflussten insbesondere das Geräuschniveau am Rücksitzplatz. In bestimmten Fällen konnte durch die Geometrie des Spoilers ein Maskierungseffekt festgestellt werden, der dazu führte, dass bestimmte Quellen durch das Mikrofonarray im Windkanal nicht erfasst wurden.

Dieses Ergebnis unterstreicht die Notwendigkeit, numerische und experimentelle Verfahren eng miteinander zu verknüpfen. Nur durch deren Kombination lassen sich verdeckte Schallquellen identifizieren und ein umfassendes Verständnis der aeroakustischen Gesamtzusammenhänge im Fahrzeug erlangen. Die in dieser Arbeit entwickelte Methodik liefert hierfür einen strukturierten und übertragbaren Ansatz, der sowohl für zukünftige Forschungsarbeiten als auch für industrielle Anwendungen von hoher Relevanz ist.

Abstract

The automotive industry faces increasing demands to ensure passenger safety and comfort, with noise reduction playing a significant role. At speeds above 100 km/h, aerodynamic noise becomes a dominant factor, driven by the turbulent flow field around the vehicle. This challenge is particularly significant for hybrid and electric vehicles, where traditional powertrain noise is diminished, amplifying the importance of aeroacoustic optimization. Computational Aeroacoustics (CAA) has emerged as an essential tool in vehicle development, offering cost-effective solutions by providing detailed data and enabling early-stage investigations using virtual models. Despite its potential, practical implementation is often constrained by high computational costs, necessitating a careful balance between prediction accuracy and efficiency to ensure integration into automotive development workflows.

In this context, hybrid numerical approaches have proven effective in addressing the constraints and time demands of the automotive industry. While extensive research exists on this topic, most studies focus on simplified geometries or vehicles, leaving a gap in the detailed documentation of numerical processes, workflows, and validation methodologies for real production vehicles. Additionally, there is limited information on how these numerical models are integrated into vehicle development processes or combined with wind tunnel measurements to perform detailed and effective analyses. Research has predominantly concentrated on the side mirror area, with minimal exploration of other critical regions. In particular, the rear part of the vehicle, where aerodynamic and acoustic interactions among multiple components are complex, remains underexplored. While this region is often considered less critical in standard configurations, it can become significant in specific cases, such as vehicles with rear aerodynamic devices, sharp-edged designs, or high-speed conditions where flow separation and wake dynamics contribute to increased noise. The literature lacks comprehensive studies on both the numerical challenges of modeling these regions and the aeroacoustic phenomena they involve.

This work introduces a three-step hybrid numerical tool integrating Computational Fluid Dynamics (CFD), a Finite Element Model (FEM), and a Vibro-Acoustic Model (VAM), with each step providing the necessary input for the next. The CFD simulation, conducted using Delayed Detached-Eddy Simulation (DDES) under incompressible flow assumptions in OpenFOAM, modeled the unsteady flow field around the vehicle. The resulting velocity field was then used in the FEM step, employing Lighthill's acoustic analogy and the Acoustic Perturbation Equations (APE) to predict the exterior acoustic field. Finally, the aerodynamic noise transmission through the vehicle's windows was modeled using the total pressure fluctuations obtained from the FEM as input for the VAM. The FEM and VAM simulations were performed using Actran, completing the process for an integrated aeroacoustic analysis.

The aeroacoustic behavior of the Lamborghini Urus SUV was analyzed at a cruising speed of 140 km/h with a 0° flow angle, focusing on aerodynamic noise phenomena in the side mirror and roof spoiler regions across a frequency range of 100–5000 Hz. Numerical simulations were validated against experimental data collected in the aeroacoustic wind tunnel of the University of Stuttgart, operated by FKFS. To guarantee consistency and repeatability, measurements were conducted on two vehicles of the same model under identical conditions. The experimental setup was carefully tailored for both the vehicle exterior and interior to align with simulation conditions and enhance the reliability of the comparative analysis. The interior noise simulation assumed that the primary contribution to passenger-perceived noise is transmitted through the windows. To align with this assumption and focus exclusively on noise generated by the vehicle's geometric features, the seals and gaps between components were taped during experiments. Additionally, a specialized internal noise insulation system was implemented to suppress potential noise sources from areas outside the scope of the investigation.

The validation process was meticulously designed to align with the structure of the simulation process, ensuring each step and output was thoroughly validated. First, a detailed analysis of the velocity and pressure fields was conducted, with validation based on time-averaged quantities calculated through CFD. Next, exterior noise propagation into the far field was assessed by comparing simulated Sound Pressure Levels (SPL) with measurements from exterior microphones positioned outside the turbulent regions near the side mirror and roof spoiler wakes. Finally, the transmission of noise into the cabin through the side window was modeled and validated against interior measurements recorded by microphones placed at the driver's and rear passenger's seats. The validation process, along with the associated experimental setup, is comprehensively documented and fully replicable, making it suitable for validating any type of hybrid process, even when using alternative CFD or aero-vibro-acoustic (AVA) numerical models.

Both exterior and interior noise predictions showed strong agreement with experimental results across the entire frequency range. Minor deviations were observed at low frequencies for the far-field SPL and the interior noise at the driver's ear, each attributed to different factors. The far-field deviation stemmed from the simulation's limited capability to model low-frequency acoustic fluctuations, constrained by the size of the simulation domain used for the aeroacoustic analysis. The interior noise deviation was linked to the simplified modeling of noise transmission into the cabin, particularly the sensitivity to boundary conditions. In the simulation, the sealing and clamping systems were neglected, and the windows were assumed to be free to vibrate, contributing to these discrepancies. Further analyses revealed a high sensitivity of the simulation at low frequencies to both the physical time and the type of wind load used in the VAM as a boundary condition. Specifically, excluding hydrodynamic pressure fluctuations and relying solely on acoustic pressure fluctuations as input led to an underestimation of low-frequency cabin noise. This behavior was explained through a wavenumber analysis, which demonstrated that hydrodynamic fluctuations dominate noise transmission below 500 Hz, while acoustic contributions prevail in the range of 500 Hz to 5 kHz.

The developed simulation process delivers highly accurate results at relatively low computational cost. This robust and versatile tool effectively analyzes different vehicle regions with distinct flow conditions, enabling targeted investigations and local optimizations during vehicle development. Each step of the process offers unique contributions: interior noise predictions assess passenger perception; acoustic propagation analysis identifies noise sources; and flow field analysis enhances understanding of noise generation mechanisms and aerodynamics. The tool's efficiency is achieved through certain simplifications, such as unsteady CFD simulations (DDES) under incompressible flow assumptions. While these assumptions limit the model's capability to account for phenomena which involve the acoustic effects into the flow field, they significantly reduce computational costs, making the process practical for vehicle development. However, the CFD phase remains the computational bottleneck, requiring significant CPU resources and substantial memory for velocity field storage. Automation is essential for maintaining usability, as manual setup of CFD, FEM, and VAM models for each configuration would significantly increase preparation time. Automating the process minimizes turnaround time and maximizes the tool's effectiveness, ensuring its practicality for aeroacoustic vehicle development.

A comprehensive analysis of aeroacoustic sources was conducted through a combined evaluation of simulation predictions and wind tunnel tests. The study revealed strong agreement between the numerical tool and microphone array measurements in terms of SPL distribution near the side mirror and roof spoiler. This was further supported by CFD analysis, which identified the primary vortex structures responsible for wind noise generation.

For the side mirror, the main source of exterior noise was attributed to vorticity released near its connection to the triangle, where it interacts with solid surfaces. This region was also identified as the dominant contributor to interior noise, as demonstrated by both wind tunnel array measurements of noise contribution into the cabin and VAM predictions of noise transmission to the driver's ear.

In the roof spoiler analysis, two design configurations were examined - a standard spoiler and a sport derivative - alongside their interactions with other aerodynamic components, such as the antenna and the rear lip spoiler. In both cases, the spoilers induced a highly turbulent field on the rear window, generating the primary exterior noise sources. These sources were also found to be the main contributors to interior noise at the rear passenger's ear. Additionally, it was discovered that, in certain scenarios, the geometry of the spoiler and the directivity of the noise sources could cause a masking effect, preventing these sources from being detected by the wind tunnel microphone array.

This finding underscores the critical importance of integrating simulations with experimental analyses. The combined approach provides a more comprehensive understanding of aeroacoustic phenomena and ensures that potential oversights, such as masked sources, are identified and addressed effectively.

1 Introduction

Nowadays, the automotive industry faces stringent demands in ensuring the safety and comfort of passengers within modern vehicles. The presence of excessive noise inside the cabin not only poses a risk to the driver's concentration but also leads to discomfort for all occupants. To address these NVH challenges specifically, automotive OEMs are increasingly adopting advanced noise reduction technologies and simulation tools to refine their development processes and enhance passenger experience.

A typical road vehicle generates a multitude of noise sources, including the powertrain, tires, and wind. Normally, at speeds above ca. 100 km/h, aerodynamic noise emerges as a dominant factor [1, 2]. In this regard, the design optimization of the car's exterior becomes crucial in reducing the transmission of wind noise into the cabin, due to the turbulent flow field surrounding the vehicle. Furthermore, the growing demand for hybrid and full-electric cars underscores the increasing significance of aeroacoustic vehicle development [3].

Within this framework, Computational Aeroacoustics (CAA) has proven essential in advancing vehicle design and supporting innovation [4]. It enhances cost-efficiency in the vehicle development process by addressing the limitations of physical experiments and providing detailed data sets that are challenging to obtain through conventional measurements. Moreover, CAA facilitates early-stage investigations when only virtual models are available. While various computational methodologies have been developed over the years [5, 6], many remain constrained by high computational costs, which limit their feasibility for widespread application in the automotive industry. Achieving an optimal balance between prediction accuracy and computational efficiency is a critical requirement for car manufacturers aiming to incorporate these numerical tools into their development processes.

For these compelling reasons, a hybrid approach [7, 8, 9, 10], based on Lighthill's acoustic analogy [11, 12] and the acoustic perturbation equations (APE) [13], has demonstrated suitability in the aforementioned context. This numerical process consists of a three-steps approach. Initially, a transient Computational Fluid Dynamics (CFD) simulation is used to solve the unsteady velocity field around the vehicle, under the assumption of incompressibility.

© The Author(s), under exclusive license to
Springer Fachmedien Wiesbaden GmbH, part of Springer Nature 2026
C. A. Perugini, *An Efficient Hybrid Computational Process for Aeroacoustic
Vehicle Development Based on Navier-Stokes Equations and Lighthill's
Acoustic Analogy*, Wissenschaftliche Reihe Fahrzeugtechnik Universität
Stuttgart, https://doi.org/10.1007/978-3-658-51977-3_1

Subsequently, a finite element (FE) model is deployed to characterize the exterior acoustic field, while a vibro-acoustic (VA) model is used to predict the noise transmitted into the cabin.

In the present work, a newly implemented hybrid numerical tool developed collaboratively by FKFS and Automobili Lamborghini S.p.A. is presented [14, 15, 16], aimed at the comprehensive characterization of wind noise sources and their correlation with interior noise within road vehicles. This efficient process, which integrates CFD simulations with acoustic models, enables engineers to assess and optimize vehicle designs early in development, reducing costs and enhancing geometric optimization. This study further seeks to validate this hybrid numerical process for aeroacoustic vehicle development, highlighting the advantages and criticalities of simulations and experiments throughout all validation phases. It provides a robust validation method applicable to various vehicle areas and components, adaptable to different solvers and numerical approaches. Additionally, the impact of a roof spoiler on exterior and interior noise is investigated, establishing a validation methodology, addressing Computational Aeroacoustics and experimental criticalities, and examining relevant acoustic phenomena. This comprehensive approach enables the evaluation of different design configurations in vehicle development, contributing to improved overall comfort and performance.

In the following sections, the aeroacoustic predictions achieved by using the above-mentioned simulation tool are presented. The SUV Lamborghini Urus is used as the reference vehicle for the entire investigation. All simulations, along with corresponding wind tunnel experiments, were conducted at a wind speed of 140 km/h. The results, presented in a frequency range up to 5kHz, include a comprehensive investigation of the regions surrounding the side mirror and the roof spoiler, respectively. The research explores the impacts of acoustic and hydrodynamic fluctuations on the side and rear windows and the resulting noise transmission phenomena. Furthermore, the study emphasizes the correlation between the primary exterior noise sources and the sound pressure level (SPL) at the passenger's ear through a joint analysis involving wind tunnel measurements and the aforementioned simulation tool.

2 Theoretical Background

For a comprehensive understanding of the simulation tool developed in this study, it is essential to delve into the theoretical background. The fundamentals of CFD based on Navier-Stokes equations are outlined. This includes hypotheses and modeling techniques, with a particular emphasis on Delayed Detached Eddy Simulation (DDES). Following this, the focus turns to CAA, laying the groundwork for an in-depth analysis of the hybrid approach proposed in this work. This covers basics of acoustics, along with the theory of Lighthill's analogy and acoustic perturbation equations, upon which the numerical process is formulated.

2.1 Computational Fluid Dynamics based on Navier-Stokes equations

The first step of the simulation model is based on an incompressible CFD simulation performed by means of a Navier-Stokes solver. Accordingly, an overview of the fundamental concepts for the fluid dynamic simulation presented later in this work follows.

2.1.1 The equations of fluid motion

Typically, conventional road vehicles operate at speeds below 200 km/h. When examining the related Mach number, defined as:

$$Ma = \frac{U_\infty}{c} \qquad \text{Eq. 2.1}$$

where U_∞ is the free stream velocity, and c the speed of sound, it usually falls below 0.3. In these operative conditions, termed as low Mach number regime, it becomes feasible to assume incompressibility of the flow field. This assumption implies treating the fluid density (ρ) as a constant, as compressibility effects are deemed negligible. Specifically, for an incompressible fluid it is possible to derive a simplified formulation of the continuity and the Navier-Stokes equations (NSE) that describe the flow field:

$$\nabla \cdot \vec{U} = 0 \qquad\qquad \text{Eq. 2.2}$$

$$\frac{\partial \vec{U}}{\partial t} + (\vec{U} \cdot \nabla)\vec{U} = -\frac{1}{\rho}\nabla p + \nu\nabla^2\vec{U} \qquad\qquad \text{Eq. 2.3}$$

Here, $\vec{U}$ represents the velocity vector field, p is the pressure, ρ the fluid density (assumed constant for incompressible flows), ν the kinematic viscosity and ∇ denotes the gradient operator. The Eq. 2.2 and Eq. 2.3 are respectively derived from the conservation of mass and momentum, extensively detailed in [1, 17, 18, 19, 20].

Computational Fluid Dynamics plays a crucial role in the study of fluid dynamics, with the Navier-Stokes equations serving as the foundation for many numerical applications. In CFD, these equations are discretized and numerically solved using the Finite Volume Method (FVM) [21, 22] to predict and visualize fluid flow behaviors. Simulations incorporating the incompressible form of the Navier-Stokes equations, such as the numerical tool presented in this work, are particularly relevant for scenarios involving low Mach numbers, such as the aerodynamics of road vehicles at typical speeds. The assumption of fluid incompressibility simplifies the governing equations, enabling computationally efficient simulations while maintaining a high level of accuracy in predicting flow fields in practical applications.

2.1.2   Navier-Stokes numerical methods

Numerous computational methodologies based on Navier-Stokes equations are currently employed within the industrial framework to tackle the fluid dynamics challenges. To gain a comprehensive understanding of the numerical tools object of this research, it is essential to provide a comprehensive overview of numerical methods commonly in use and delve into the mathematical models that underlie the modeling methodologies. The most relevant numerical methods based on NSE can be summarized as follows:

- Reynolds Averaged Navier Stokes (RANS) simulations: This approach constitutes widely adopted methodologies in industrial fluid dynamics, offering a balance between computational efficiency and accuracy. RANS models provide time-averaged solutions, making them suitable for steady-state problems. Unsteady Reynolds Averaged Navier Stokes (URANS) extends this approach by incorporating temporal variations, allowing for the simulation of unsteady flows with improved accuracy compared to RANS. These methods find extensive use in automotive aerodynamics, combustion modeling, and heat transfer analyses. The suitability to capture mean flow patterns and turbulence characteristics makes RANS and URANS simulations valuable tools for predicting overall vehicle performance and assessing design modifications.

- Large-Eddies and Detached-Eddies Simulations (LES and DES): These methods offer a transition towards more detailed and unsteady flow analyses. LES resolves larger turbulent structures explicitly, capturing the energetic eddies, while modeling the smaller scales. DES combines elements of RANS and LES, using RANS in attached boundary layers and transitioning to LES in separated or transitional flow regions. LES and DES simulations are particularly advantageous in scenarios where unsteady and separated flow phenomena play a critical role. Automotive applications include the study of wake turbulence, flow separation around vehicle components, the prediction of unsteady forces that influence vehicle stability and aerodynamic performance.

- Direct Numerical Simulation (DNS): It represents the utmost fidelity in computational fluid dynamics. Unlike RANS, URANS, LES, and DES, DNS solves the Navier-Stokes equations without any turbulence modeling, providing a detailed representation of all flow scales. However, the computational cost associated with resolving the entire turbulent spectrum limits DNS application to fundamental research and academic studies.

To grasp the fundamentals of RANS simulations, and consequently LES and DES, it is necessary to introduce the Reynolds decomposition [23]. This decomposition involves breaking down turbulent quantities, such as local velocity components (u, v, w) and pressure (p), into mean and fluctuating components. The decomposition of the above-mentioned quantities can be expressed as follows:

$$u(t) = \bar{u} + u'(t)$$
$$v(t) = \bar{v} + v'(t)$$
$$w(t) = \bar{w} + w'(t)$$
$$p(t) = \bar{p} + p'(t)$$

Eq. 2.4

$$\bar{u} = \frac{1}{\Delta t} \int_{t1}^{t2} u(t)dt$$

Eq. 2.5

Here, $\bar{u}$, $\bar{v}$, $\bar{w}$ and $\bar{p}$ represent the time-averaged independent term, while u'(t), v'(t), w'(t) and p'(t) are defined as the fluctuating terms. Taking the x-component of velocity as an example, the temporal mean value ($\bar{u}$) can be calculated as shown in Eq. 2.5, while the fluctuating component as the difference between the instantaneous value (u(t)) and the mean term ($\bar{u}$). It must be noted that the time interval (Δt) must be chosen such that the mean value is independent of the averaging interval. A graphical representation of the Reynolds decomposition and the related terms is provided in **Figure 2.1**.

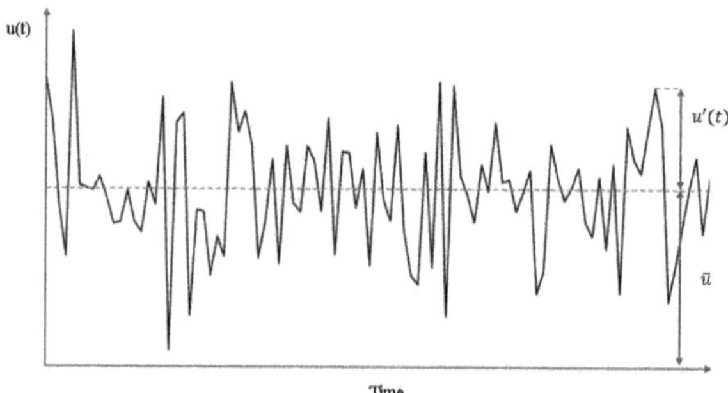

Figure 2.1: Illustration of Reynolds decomposition for the x-component of the velocity (u(t))

Considering the Reynolds decomposition of each variable (Eq. 2.5) in the Eq. 2.2 and Eq. 2.3, and averaging over time, we obtain the following set of four scalar equations known as the Reynolds-Averaged continuity and momentum equations respectively:

$$\frac{\partial \bar{u}}{\partial x} + \frac{\partial \bar{v}}{\partial y} + \frac{\partial \bar{w}}{\partial z} = 0 \qquad\qquad \text{Eq. 2.6}$$

$$\frac{\partial \bar{u}}{\partial t} + \bar{u}\frac{\partial \bar{u}}{\partial x} + \bar{v}\frac{\partial \bar{u}}{\partial y} + \bar{w}\frac{\partial \bar{u}}{\partial z}$$
$$= -\frac{1}{\rho}\frac{\partial \bar{p}}{\partial x} + v\left(\frac{\partial^2 \bar{u}}{\partial x^2} + \frac{\partial^2 \bar{u}}{\partial y^2} + \frac{\partial^2 \bar{u}}{\partial z^2}\right)$$
$$- \left(\frac{\partial \overline{u'^2}}{\partial x} + \frac{\partial \overline{u'v'}}{\partial y} + \frac{\partial \overline{u'w'}}{\partial z}\right)$$

$$\frac{\partial \bar{v}}{\partial t} + \bar{u}\frac{\partial \bar{v}}{\partial x} + \bar{v}\frac{\partial \bar{v}}{\partial y} + \bar{w}\frac{\partial \bar{v}}{\partial z}$$
$$= -\frac{1}{\rho}\frac{\partial \bar{p}}{\partial y} + v\left(\frac{\partial^2 \bar{v}}{\partial x^2} + \frac{\partial^2 \bar{v}}{\partial y^2} + \frac{\partial^2 \bar{v}}{\partial z^2}\right) \qquad \text{Eq. 2.7}$$
$$- \left(\frac{\partial \overline{u'v'}}{\partial x} + \frac{\partial \overline{v'^2}}{\partial y} + \frac{\partial \overline{v'w'}}{\partial z}\right)$$

$$\frac{\partial \bar{w}}{\partial t} + \bar{u}\frac{\partial \bar{w}}{\partial x} + \bar{v}\frac{\partial \bar{w}}{\partial y} + \bar{w}\frac{\partial \bar{w}}{\partial z}$$
$$= -\frac{1}{\rho}\frac{\partial \bar{p}}{\partial z} + v\left(\frac{\partial^2 \bar{w}}{\partial x^2} + \frac{\partial^2 \bar{w}}{\partial y^2} + \frac{\partial^2 \bar{w}}{\partial z^2}\right)$$
$$- \left(\frac{\partial \overline{u'w'}}{\partial x} + \frac{\partial \overline{v'w'}}{\partial y} + \frac{\partial \overline{w'^2}}{\partial z}\right)$$

For steady-state analysis, the time derivative term (i.e., the first of the Eq. 2.7) can be neglected. In such instances, the equations are referred to as Reynolds-Averaged Navier-Stokes (RANS). Otherwise, when the time derivative term is retained, they are termed Unsteady RANS (URANS). In both cases, the right-hand side is characterized by an additional force term which is the divergence of the Reynolds stress tensor ($\tau'_{i,j}$). The tensor is defined as follows:

$$\tau'_{i,j} = -\rho\overline{U'_i U'_j} = -\rho \cdot \begin{bmatrix} \overline{u'^2} & \overline{u'v'} & \overline{u'w'} \\ \overline{u'v'} & \overline{v'^2} & \overline{v'w'} \\ \overline{u'w'} & \overline{v'w'} & \overline{w'^2} \end{bmatrix} \qquad \text{Eq. 2.8}$$

This term represents the correlation between the fluctuating velocity components in different directions and accounts for the effects of turbulence stresses on the mean flow. In other words, it quantifies the transport of momentum due to turbulent fluctuations [24].

2.1.3   The closure problem and overview of turbulence models

The Reynolds stress tensor, resulting from temporal averaging, is a key term in the RANS equations. It allows for the inclusion of turbulence effects without the need to directly resolve all turbulent fluctuations. As a symmetric tensor, it is characterized by six independent components. These components represent additional unknowns in our system of equations, alongside pressure and the velocity field components, bringing the total number of unknowns to ten. Given that the equations describing fluid motion consist of four scalar equations (i.e. continuity and momentum equations), the system is underdetermined, and the Reynolds stress tensor cannot be directly calculated. Accurate modeling of this tensor is essential for predicting the behavior of turbulent flows. Therefore, turbulence models are commonly used in Computational Fluid Dynamics (CFD) solvers to provide additional equations, thereby closing the system of equations [20].

An overview of the main turbulence models commonly used is depicted in the following figure:

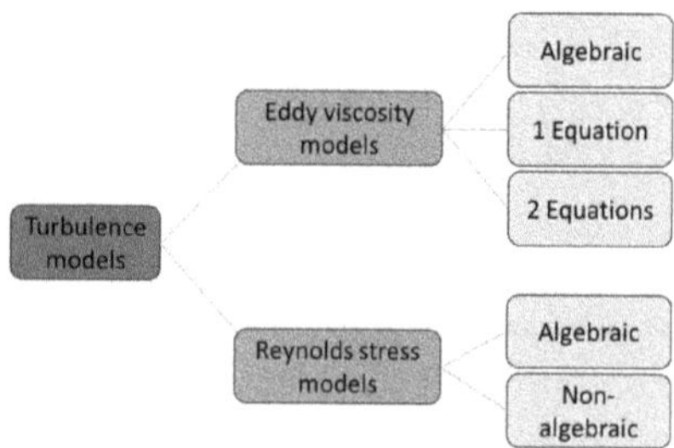

Figure 2.2:   Overview on turbulence models [1]

We can categorize turbulence models into two main types: eddy viscosity models (EVMs) and Reynolds stress models (RSMs). EVMs model turbulent stresses by assuming isotropic turbulence and using the concept of eddy viscosity (v_t), which relates the turbulent stresses to the mean velocity gradient tensor through a turbulent viscosity coefficient [24, 25]. In most EVMs, this relationship is linear, with the traceless part of the Reynolds stress tensor being proportional to the mean strain tensor, and the eddy viscosity serving as the proportionality factor. These models can be further subdivided into algebraic and differential approaches. Algebraic models solve a direct relationship between turbulent quantities without introducing additional transport equations (e.g., zero-equation models), while non-algebraic models, such as one-equation and two-equation models (e.g., the Spalart-Allmaras, k-ω and k-ε models), involve solving additional partial differential equations for turbulence properties like turbulent kinetic energy (TKE) or dissipation rate (ε). The latter were employed in the numerical analyses conducted in this study due to their computational efficiency, robustness, and broad applicability in various engineering problems.

In contrast, RSMs solve transport equations for each component of the Reynolds stress tensor, offering a more detailed and accurate representation of turbulence. Unlike EVMs, which use a scalar eddy viscosity to model turbulence effects, RSMs directly calculate the Reynolds stresses, accounting for the anisotropy of turbulence. This approach enables RSMs to capture the directional effects of turbulence and provides improved predictions for complex flows characterized by significant anisotropy, swirl, and separation. While RSMs deliver greater accuracy and a more comprehensive description of turbulence, they come with increased computational cost and complexity.

2.1.4   Near wall modeling

A critical aspect of turbulence prediction is the behavior of the flow near solid boundaries, especially in turbulent regimes. This behavior is governed by two fundamental conditions: the non-permeable condition and the no-slip condition:

- **Non-Permeable condition**: This condition implies that there is no flow of fluid through the surface of the solid boundary. Mathematically, it is expressed as:

$$\vec{U} \cdot \hat{n} = 0 \qquad\qquad \text{Eq. 2.9}$$

where $\hat{n}$ is the unit normal vector to the surface.

- **No-Slip condition**: the fluid velocity at the solid surface is equal to the velocity of the surface itself. For a stationary boundary, this means that the fluid velocity at the surface is zero:

$$\vec{U}|_{wall} = 0 \qquad\qquad \text{Eq. 2.10}$$

As we move farther from the boundary, the interaction between the flow and the solid surface generates what is known as the boundary layer. The thickness of this region, denoted as δ, is defined as the perpendicular distance from the wall to the point where the flow velocity reaches 99% of the free stream velocity ($U = 0.99U_\infty$). To better understand and describe the flow field behavior in this area, the following non-dimensional quantities must be defined:

$$u^+ = \frac{U}{U_\tau} \qquad\qquad \text{Eq. 2.11}$$

$$y^+ = \frac{U_\tau \cdot d}{\nu} \qquad\qquad \text{Eq. 2.12}$$

Here, u^+ is the dimensionless velocity, defined as the velocity scaled by the friction velocity U_τ, defined in Eq. 2.13 (being τ_w the wall shear stress and ρ the fluid density). y^+ is the dimensionless wall distance, defined as the distance from the wall (d) scaled by the kinematic viscosity (ν) and the friction velocity (U_τ).

$$U_\tau = \sqrt{\frac{\tau_w}{\rho}} \qquad \text{Eq. 2.13}$$

According to the boundary layer theory [26, 27], this region can be divided into three main sub-regions:

1. **Viscous sublayer:** For $y^+ < 5$ the flow is dominated by viscous shear stress, and the velocity profile is approximately linear. The inner layer law can be expressed as:

$$u^+ = y^+ \qquad \text{Eq. 2.14}$$

2. **Buffer region:** This transitional region bridges the viscous sublayer and the log-law region. In this layer, both viscous and turbulent effects are significant, and the velocity profile transitions from linear to logarithmic. The buffer layer typically corresponds to $5 < y^+ < 30$.

3. **Logarithmic layer:** Further from the wall, for $30 < y^+ < 300$, the flow becomes fully turbulent, and the velocity profile follows a logarithmic distribution [26]:

$$u^+ = \frac{1}{\kappa}\ln(y^+) + B \qquad \text{Eq. 2.15}$$

Here, κ is the von Kármán constant (approximately 0.41), and B is an empirical constant (approximately 5.5).

In **Figure 2.3**, a schematic representation of the boundary layer including main regimes (laminar and turbulent) and the related sub-regions is depicted. In **Figure 2.4**, an example of an experimental velocity profile in a turbulent boundary layer is compared to the viscous sublayer and the logarithmic law.

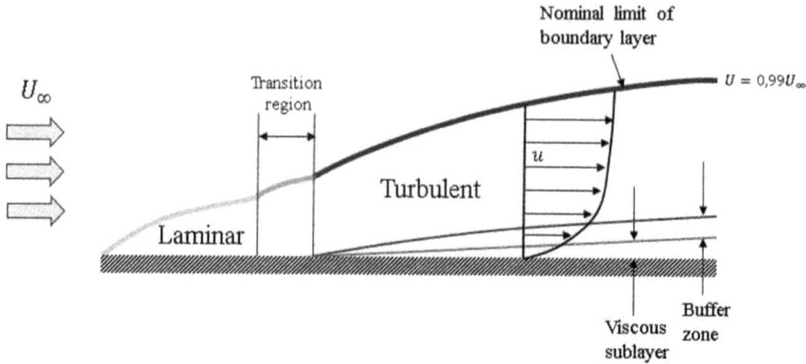

Figure 2.3:   Schematic representation of boundary layer and the related regions

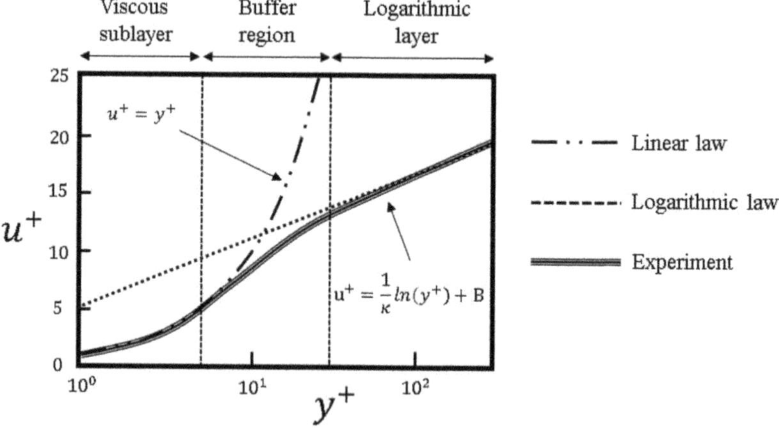

Figure 2.4:   Example of experimental velocity profile in the different regions of a turbulent boundary layer [26]

To accurately resolve the boundary layer, two primary approaches are typically employed:

- **Low-Reynolds (L-R) approach**: these models attempt to resolve the entire boundary layer, including the viscous sublayer, buffer layer, and logarithmic layer. This approach requires a very fine mesh near the wall, with y^+ values typically lower than 1, see **Figure 2.5** left.

- **High-Reynolds (H-R) approach**: these models use empirical functions to predict the near-wall region, reducing the need for extremely fine meshes. Wall functions are typically used when the mesh resolution is insufficient to resolve the viscous sublayer directly. The standard wall function approach assumes that the first cell center is in the logarithmic layer (typically $30 < y^+ < 300$), see **Figure 2.5** right.

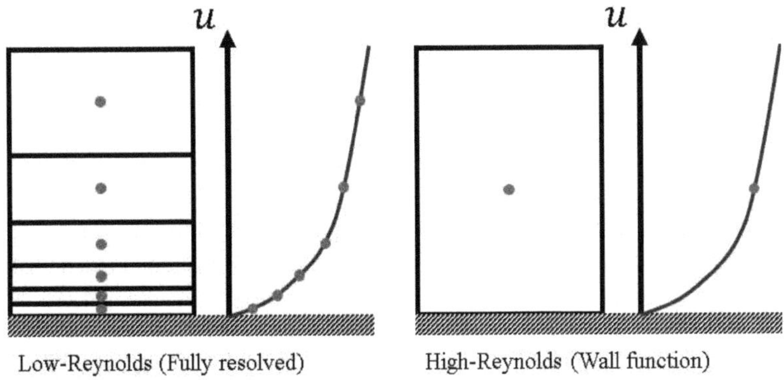

Figure 2.5:   Boundary layer discretization strategy for Low-Reynolds and High-Reynolds approach

The contribution of boundary layer modeling to the overall computational cost is significant. In this context, the H-R approach has proven to be more computationally efficient than the L-R approach. This efficiency is due to the H-R approach's use of a wall function and a simplified mesh, making it more suitable for industrial applications where computational efficiency and cost-effectiveness are critical [28, 29].

2.1.5   Delayed Detached Eddies Simulations with Spalart-Allmaras turbulence model

In order to address the critical issue of boundary layer modeling while managing computational costs effectively, a hybrid simulation model (i.e., DES) was first proposed in 1997 [30]. It was developed, to combine the computational efficiency of RANS with the accuracy of LES.

RANS models are effective in solving for the mean flow properties by averaging the Navier-Stokes equations. This approach, relying on the Reynolds decomposition (see section 2.1.2), focuses on the mean velocity and mean pressure, thus efficiently capturing steady-state flows but often at the expense of detailed turbulence characteristics. Conversely, LES directly resolves the large-scale turbulent structures and models the smaller scales using a subgrid-scale (SGS) model. LES employs a spatial filtering operation, decomposing the velocity field components into resolved and subgrid-scale components. For instance, applying the spatial filtering to the velocity x-component leads to:

$$u = \tilde{u} + u_{SGS}$$

Eq. 2.17

where $\tilde{u}$ represents the resolved scales and u_{SGS} the subgrid scales. This method provides a more detailed representation of turbulence but at a significantly higher computational cost [24]. Further details ca be found in [31, 25].

DES uses the RANS approach in boundary layer regions and switches to LES in external flow field and in separated flow regions. This hybrid approach significantly reduces the number of grid points and time steps needed compared to a full LES, thereby reducing the computational cost without compromising the accuracy of large-scale turbulent structures [28]. Specifically, it aims to optimally blend RANS and LES by dynamically switching between the two based on the local flow conditions.

An integral part of the DES framework is the Spalart-Allmaras turbulence model, which is often employed, including this study, for the RANS portion of DES due to its simplicity and robustness. The Spalart-Allmaras model is a one-equation model (see section 2.1.3) that solves a transport equation for a modified eddy viscosity ($\tilde{v}$). Its original formulation [32] is given by:

$$\frac{\partial \tilde{v}}{\partial t} + \vec{U} \cdot \nabla \tilde{v} = C_{b1}\tilde{S}\tilde{v} + \frac{1}{\sigma}\left[\nabla \cdot \left((v + \tilde{v})\nabla \tilde{v}\right) + C_{b2}(\nabla \tilde{v})^2\right] \\ - C_{w1}f_w\left(\frac{\tilde{v}}{d}\right)^2$$

Eq. 2.18

where:

- $\tilde{v}$ is the modified turbulence viscosity,
- $\vec{U}$ is the velocity vector,
- $\tilde{S}$ intermediate variable, function of the vorticity magnitude (S) and $\tilde{v}$,
- C_{b1}, C_{b2}, C_{w1}, σ are model constants,
- f_w is the damping functions,
- d is the distance to the nearest wall.

In DES, the S-A model is adapted to switch between RANS and LES regions. The key modification involves replacing the distance to the nearest wall (d) with a modified length scale (d_{DES}), defined as:

$$d_{DES} = min(d, \ C_{DES}\Delta)$$

Eq. 2.19

Where Δ is the largest of the grid spacing in all the three directions (as defined in Eq. 2.20), and C_{DES} ($= 0.65$) an empirical constant:

$$\Delta \equiv \ max\,(\Delta x, \Delta y, \Delta z)$$

Eq. 2.20

This substitution allows the model to dynamically adjust the turbulence treatment based on the local grid resolution. The S-A model equation in DES (known as DES97 model) thus becomes:

$$\frac{\partial \tilde{v}}{\partial t} + \vec{U} \cdot \nabla \tilde{v} = \ C_{b1}\tilde{S}\tilde{v} + \frac{1}{\sigma}\left[\nabla \cdot \left((v + \tilde{v})\nabla \tilde{v}\right) + C_{b2}(\nabla \tilde{v})^2\right]$$
$$- C_{w1}f_w\left(\frac{\tilde{v}}{d_{DES}}\right)^2$$

Eq. 2.21

As a result, in boundary layers where the condition $d_{DES} \ll \Delta$ is met, RANS model is employed. Conversely, on a grid that is appropriate for resolving eddies generated after significant separation, we typically find $\Delta \ll d_{DES}$, and LES model is used. This leads to a single model functioning as a S-A model under the first condition and as an SGS model under the second. It is important to note that if there is finer resolution in one direction, it does not affect d_{DES}

so the smallest resolved eddies still scale with Δ, and the additional resolution is essentially unused [30].

By combining the Spalart-Allmaras model with LES in the DES framework, the method achieves a balance between efficiency and accuracy. The RANS mode efficiently handles near-wall and attached flow regions, while the LES mode captures the intricate dynamics of separated flows and large-scale turbulence. This hybrid approach is particularly useful in simulating flows over complex geometries where both attached and separated flow regions are present, such as in automotive applications.

Despite its advantages, DES have some limitations [32], particularly near the transition zones between RANS and LES. The abrupt switch can sometimes lead to grid-induced separation, resulting in inaccuracies. To address this, the Delayed Detached Eddy Simulation (DDES) was proposed as an improvement [33, 34]. DDES introduces a shielding function that delays the transition from RANS to LES in the near-wall regions, preventing premature switching that could lead to erroneous results. This is achieved by modifying the distance to the wall d with a new length scale $\tilde{d}_{DES}$, defined as:

$$\tilde{d}_{DES} = d - f_d max(0 \, ; \, d - C_{DES}\Delta \,) \qquad\qquad \text{Eq. 2.22}$$

The shielding function f_d ensures that the model acts as RANS near the wall and as LES away from the wall. It is defined as:

$$f_d = 1 - \tanh[(8\, r_d)^3] \qquad\qquad \text{Eq. 2.23}$$

where r_d is a dimensionless parameter (see complete formulation in [33]). This function effectively delays the DES transition, thus mitigating the risk of grid-induced separation and improving the overall accuracy of the simulation. Notably, setting f_d to 0 results in the RANS mode ($\tilde{d}_{DES} = d$), whereas setting f_d to 1 produces the DES97 mode ($d_{DES} = min(d, \, C_{DES}\Delta \,)$).

For this study, a DDES has been developed to achieve an optimal balance between high-fidelity results and computational efficiency. By leveraging the hybrid nature of DDES, which integrates the strengths of URANS and LES techniques, this study aims to enhance the predictive capabilities of aeroacoustics models while maintaining manageable computational costs. This balance is crucial for practical applications in the vehicle development process. Numerous applications of this simulation type in the automotive field have been published over the years, demonstrating its efficacy in predicting complex flow phenomena and noise generation mechanisms.

2.1.6 Applications and limitations of DDES

Delayed Detached Eddy Simulations are applied across a wide array of fields due to its capability in accurately modeling turbulent flows. This versatile approach has proven highly effective in various sectors, with applications spanning the automotive, aerospace, and turbomachinery industries, as well as environmental engineering and wind energy systems.

For instance, DDES are widely employed in the automotive industry to enhance vehicle aerodynamics, improve aeroacoustic performance, and optimize thermal management [35, 36, 37]. In aerospace engineering, DDES help to predict and analyze complex aerodynamic phenomena around airfoils, wings, and fuselages, optimizing both aerodynamic and aeroacoustic design [38]. In turbomachinery, DDES are crucial for detailed flow analysis in turbines, compressors, and pumps, accurately predicting unsteady flow phenomena such as blade-vortex interactions and flow separation, thereby optimizing performance and efficiency [39]. Additionally, the environmental and energy sector benefit from DDES in studying atmospheric boundary layers, pollutant dispersion, and wind flow around urban structures for assessing air quality and designing effective pollution control strategies [40].

Although DDES was originally introduced to reduce computational costs while maintaining reasonable accuracy, its application still involves certain limitations. Achieving high-fidelity results requires a sufficiently refined grid, particularly in regions where the LES component is active. This necessity can lead to increased computational costs and complexity in grid generation [33]. Another limitation is the model's sensitivity to the transition between URANS and LES regions. Inaccurate modeling of this transition can result in either over-dissipation of turbulence or insufficient resolution of small-scale structures, affecting the overall accuracy of the simulation. Additionally, DDES relies on empirical constants and model functions that may not be universally applicable to all flow scenarios, necessitating careful calibration and validation for each specific case [28, 38]

Despite these challenges, the continued development and refinement of DDES (e.g.; IDDES and ZDDES) methodologies hold promise for even greater accuracy and efficiency [41]. Future research is likely to focus on improving the robustness of the model transition, enhancing grid generation techniques, and developing more universally applicable model constants. There is also ongoing work to integrate machine learning techniques to optimize grid generation and model calibration processes, which could significantly reduce the time and computational resources required for simulations [42].

2.2 Computational Aeroacoustics based on Lighthill's acoustic analogy

The second step of the simulation model utilizes Computational Aeroacoustics techniques, relying on Lighthill's acoustic analogy and the Acoustic Perturbation Equations (APE). This section provides an overview of the theoretical basis necessary for understanding the aeroacoustic analysis conducted later in this work.

2.2.1 Aeroacoustic sources and sound propagation

In the context of CAA for road vehicles, understanding the origins and mechanisms of sound generation is crucial for accurately predicting and analyzing noise produced by cars. Aeroacoustic sources, which means the origins of sound within a vehicle's fluid flow, are typically generated by the interaction of turbulent structures with the vehicle's surface or within the air itself.

The generation of aerodynamic noise is primarily attributed to three distinct mechanisms [1, 2, 43, 44, 45]:

1. **Unsteady volumetric flow**: This mechanism involves the fluctuating pressure due to volume flow variations, such as those caused by leaks in sealing systems. It is idealized as monopole sources, where the sound is

characterized by a radial pressure field p=$f(r)$, with r being the radial distance from the source. These sources dominate at low Mach regimes and are significant in leak noise scenarios, with their intensity I_m proportional to the flow velocity raised to the fourth (Eq. 2.24):

$$I_m \sim \frac{\rho}{c} U^4 = \rho \cdot Ma \cdot U^3 \qquad \text{Eq. 2.24}$$

2. **Solid wall interaction:** This mechanism occurs when unsteady pressures impact rigid surfaces, such as when vortices shed or interact with the vehicle body. These may be represented by dipole sources, which consist of two adjacent monopole sources oscillating out of phase, described by p=$f(r,\theta)$, with θ representing the directional dependence on pressure radiation. Dipole sources are prominent in scenarios involving surface interactions, with sound intensity (I_d) proportional to the flow velocity raised to the sixth power (Eq. 2.25). In the absence of leak noise, dipole sources typically dominate cabin noise.

$$I_d \sim \frac{\rho}{c^3} U^6 = \rho \cdot Ma^3 \cdot U^3 \qquad \text{Eq. 2.25}$$

3. **Turbulent shear stresses:** This noise is generated by unsteady internal stresses within a fluid, particularly in turbulent shear layers or vehicle wakes. They are modeled by quadrupole sources, which are combinations of two dipole sources. Generally, less significant in vehicle aeroacoustics due to the relatively low speeds, with sound intensity (I_q) proportional to the flow velocity raised to the eighth power (Eq. 2.26) but become relevant at higher speeds due to the intense turbulent stresses.

$$I_q \sim \frac{\rho}{c^5} U^8 = \rho \cdot Ma^5 \cdot U^3 \qquad \text{Eq. 2.26}$$

The following figure provides an idealized representation of the aforementioned aeroacoustic source types:

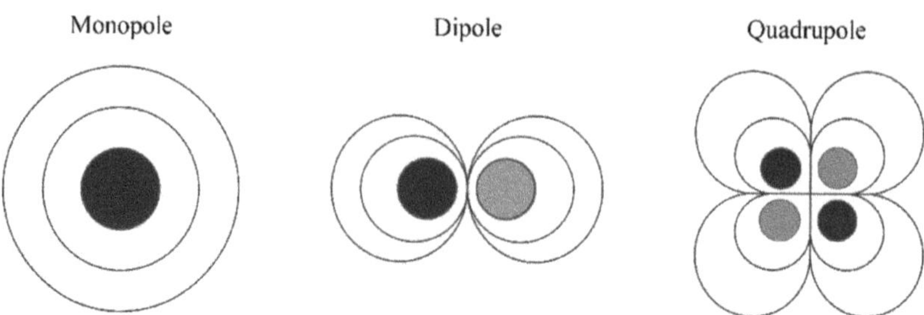

Figure 2.6: Representation of the aeroacoustic source types

In this context, it is also useful to introduce the definitions of sound pressure and sound pressure level. Sound pressure represents the local deviation from atmospheric pressure due to the presence of sound waves and is measured in Pascal (Pa). Human hearing can detect a vast range of sound pressures, from the threshold of hearing around $2\text{x}10^{-5}$ Pa, to the threshold of pain at approximately $2\text{x}10^{2}$ Pa. Because this range spans several orders of magnitude, a logarithmic scale is used to quantify sound pressure, known as the sound pressure level (SPL). SPL is calculated as the logarithmic ratio of a given RMS (root mean square) sound pressure p_{rms} to a reference pressure p_0, which is typically the hearing threshold ($2\text{x}10^{-5}$ Pa). This relationship is expressed in decibels (dB) as follows:

$$SPL = 20 \log(\frac{p_{rms}}{p_0}) \qquad\qquad \text{Eq. 2.27}$$

When dealing with multiple sound sources, it is important to note that their SPL values do not simply combine linearly due to the logarithmic nature of SPL. Instead, the individual sound pressures must be combined linearly before converting back to SPL. This is particularly relevant in environments with multiple incoherent noise sources, such as the interior of a vehicle, where different sound contributions merge to form the overall sound level.

2.2.2   Generation mechanisms of wind noise around a vehicle

After introducing the source types in the previous chapter, it is crucial to understand how these sources are typically generated around a road vehicle. The most relevant geometry features in this sense include the A-pillar, door mirrors, windscreen wipers, underfloor, body sealing, and open cavities [1, 2]. An example of the potential sources and the related fluid phenomena generated around the driver area is depicted in **Figure 2.7**.

Monopole sources (**Figure 2.6**, left) can occur in specific situations such as leaks in the car seal system or between component gaps (e.g., doors gap, see unsteady volumetric flow in **Figure 2.7**), creating pulsating sounds that radiate equally in all directions. Preventing these monopole sources is essential for reducing unwanted noise and enhancing the overall acoustic comfort within the vehicle.

Dipole sources (**Figure 2.6**, center) are typically produced by the interaction of airflow with side mirrors, windshield wipers, and A-pillar (see solid surface interaction in **Figure 2.7**). For instance, the aerodynamic noise from side mirrors is a significant dipole source, resulting from the unsteady aerodynamic forces acting on the mirror surface. This type of noise usually significantly impacts the acoustics inside the vehicle cabin potentially disturbing the car's passengers.

Quadrupole sources (**Figure 2.6**, right) are generally generated in the wake behind the vehicle or by appendices such as the side mirror (see turbulent shear stresses in **Figure 2.7**). These sources become significant in high-speed conditions where the turbulence intensity is greater. The sound radiation from quadrupole sources is generally more complex and diffused, often contributing to the overall noise in high-speed driving conditions.

Once generated, sound waves induce pressure fluctuations on the car surface which are potentially transmitted inside the vehicle cabin. Thus, a comprehensive understanding of the sound field is essential in the context of vehicle development to employ correct measurement setup and to effectively investigate aeroacoustic phenomena.

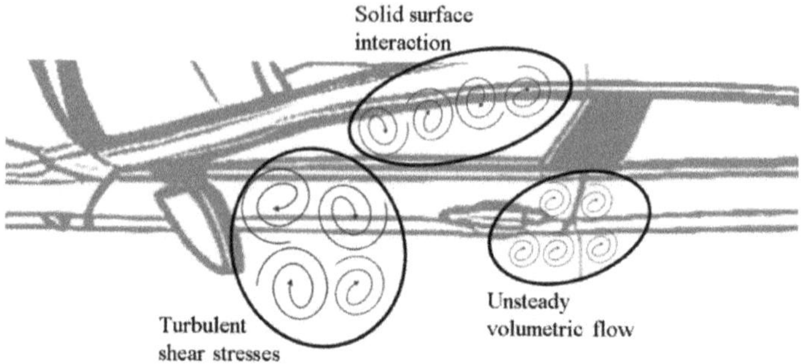

Figure 2.7:   Example of the different source type that are typically generated around a vehicle [2]

In road vehicle aeroacoustics, reflection plays a significant role in shaping the sound field around and within the vehicle. Sound waves reflect off the vehicle's surfaces and other nearby obstacles, potentially creating interference patterns that can influence perceived noise levels, particularly inside the cabin. These interactions are critical in understanding and managing noise hotspots that may arise due to the vehicle's geometry. Absorption is another important phenomenon, particularly within the vehicle interior, where materials like upholstery and soundproofing layers attenuate sound waves and contribute to noise reduction. Other effects such as refraction, diffraction, and scattering by turbulence are generally negligible in the context of vehicle aeroacoustics and are typically excluded from standard analyses, as they have a minor influence in the near-field region of the vehicle.

2.2.3   Origins of the Lighthill's analogy and implementation in a finite element framework

In the context of computational aeroacoustics, Lighthill's analogy, introduced by Michael James Lighthill in 1952 [11, 12], represents a significant advancement in modeling sound generation by turbulent fluid flows. This analogy reformulates the complex Navier-Stokes equations, specifically tailored for acoustics analysis. In this section, the mathematical formulation of Lighthill's analogy, its derivation from the Navier-Stokes equations, and the potential implementation in a finite element framework are explored.

The Navier-Stokes equations, previously discussed in chapter 2.1.1 for an incompressible fluid under no external forces, can be generalized to include compressibility effects [1, 8, 17]. For the sake of clarity in the mathematical discussion, we will use indicial notation in the following formulation:

$$\frac{\partial \rho}{\partial t} + \frac{\partial (\rho u_i)}{\partial x_i} = 0 \qquad \text{Eq. 2.28}$$

$$\frac{\partial (\rho u_i)}{\partial t} + \frac{\partial (\rho u_i u_j)}{\partial x_j} = -\frac{\partial p}{\partial x_i} + \frac{\partial \tau_{ij}}{\partial x_j} \qquad \text{Eq. 2.29}$$

Here, τ_{ij} is the viscous stress tensor defined as for a Newtonian fluid:

$$\tau_{ij} = \mu \left(\frac{\partial u_i}{\partial x_j} + \frac{\partial u_j}{\partial x_i} - \frac{2}{3} \frac{\partial u_k}{\partial x_k} \delta_{ij} \right) \qquad \text{Eq. 2.30}$$

With μ defined as dynamic viscosity of the fluid and δ_{ij} as the Kronecker delta. By transforming Eq. 2.28 and Eq. 2.29 [11, 46, 47], we can derive a wave equation representing the well-known Lighthill's acoustic analogy:

$$\frac{\partial^2}{\partial t^2} (\rho - \rho_0) - c_0^2 \frac{\partial^2}{\partial x_i \partial x_i} (\rho - \rho_0) = \frac{\partial^2 T_{ij}}{\partial x_i \partial x_j} \qquad \text{Eq. 2.31}$$

In this equation, the left-hand side (LHS) consists of an acoustic wave operator, where ρ_0 and c_0 are the reference density and reference speed of sound in a medium at rest, respectively. The right-hand side (RHS) is the source term dependent on Lighthill's stress tensor (T_{ij}), defined as follows:

$$T_{ij} = \rho u_i u_j + (p - c_0^2 \rho) \delta_{ij} - \tau_{ij} \qquad \text{Eq. 2.32}$$

In low Mach regime and high-Reynolds number flows, the compressibility effects of the fluid are minimal, meaning that density variations due to pressure fluctuations may be neglected [8, 48, 49]. As a result, the pressure field does not significantly affect the velocity field, allowing the flow to be treated as incompressible. Under these conditions, the following assumptions are typically made:

- **Inviscid flow**: Neglecting viscosity, so $\tau_{ij} \approx 0$.
- **Isentropic process**: No entropy changes, so the equation of state simplifies to a constant sound speed c.
- **Negligible pressure deviations**: Assuming $(p - c_0^2 \rho) \approx 0$, meaning pressure deviations are small or uniform.

With these assumptions, Lighthill's stress tensor simplifies to:

$$T_{ij} = \rho u_i u_j \qquad\qquad \text{Eq. 2.33}$$

Thus, Lighthill's analogy can be rewritten in the following simplified form:

$$\frac{\partial^2}{\partial t^2}(\rho - \rho_0) - c_0^2 \frac{\partial^2}{\partial x_i \partial x_i}(\rho - \rho_0) = \frac{\partial^2 \rho u_i u_j}{\partial x_i \partial x_j} \qquad\qquad \text{Eq. 2.34}$$

This form emphasizes the contribution of Reynolds stress to sound generation, neglecting the pressure deviation term. This approximation is valid in cases where the flow is dominated by turbulent motion, and pressure variations are negligible compared to turbulent stresses.

In this scenario, the source term is expressed solely as a function of velocity fluctuations. This clearly shows how using an unsteady incompressible CFD simulation. We can employ the calculated transient velocity field as an input for Lighthill's analogy to calculate potential aeroacoustic sources. To do so, the variational formulation of Lighthill's analogy (Eq. 2.35), originally proposed by Oberai [8, 50, 51, 52], is frequently applied in this context. Notably, this formulation is integrated into the Actran software, which was used in this study.

$$\int_{\Omega} \left(\frac{\partial^2}{\partial t^2}(\rho - \rho_0)\delta\rho + c_0^2 \frac{\partial}{\partial x_i}(\rho - \rho_0)\frac{\partial \delta\rho}{\partial x_i} + \frac{\partial T_{ij}}{\partial x_j}\frac{\partial \delta\rho}{\partial x_i} \right) dx$$
$$= 0$$

Eq. 2.35

This variational formulation adapts the acoustic wave equation for finite element methods, ensuring it can be numerically simulated by integrating the terms over an arbitrary computational domain Ω. In this context, a test function $\delta\rho$ is introduced, which is standard in variational formulations to guarantee the equation's validity for any arbitrary density perturbation. The formulation includes three key components:

- The second time derivative of the density perturbation
- The spatial variations of the density perturbation
- The effects of the Lighthill stress tensor T_{ij}, which includes contributions from turbulent stresses

By incorporating these elements, the variational formulation enables accurate and robust simulations of acoustic phenomena using finite element methods.

2.2.4   Origins of the Acoustic Perturbation Equations

In fluid dynamics and aeroacoustics, understanding the interplay between acoustic and hydrodynamic fields is crucial for accurately predicting sound generation and propagation. The acoustic perturbation equations (APEs) provide a framework to separate the acoustic contributions from the hydrodynamic components within a fluid flow.

The fluctuations in the flow are characterized by their associated length scales. Acoustic wavelengths (λ) are large and propagate fast at the speed of sound, while hydrodynamic fluctuations are characterized by smaller scales (l) and propagate at the relatively low local flow velocity. For a specific frequency (f), these scales are given by the following formulas:

$$\lambda = \frac{c}{f}$$

Eq. 2.36

$$l = \frac{U}{f} \qquad\qquad \text{Eq. 2.37}$$

The derivation of Lighthill's acoustic analogy from the compressible Navier-Stokes equations encompasses all length scales without simplifications. However, in the far-field region, hydrodynamic fluctuations dissipate quickly, leaving only acoustic fluctuations in the pressure field. For analytical clarity, it is useful to exclude the hydrodynamic pressure components. Ewert et al. [13] proposed a method that filters out hydrodynamic contributions by isolating acoustic modes based on their different propagation speeds compared to hydrodynamic modes. The equations are thus modified to focus solely on aeroacoustic sources [52, 53]:

$$\frac{\partial^2}{\partial t^2}(\rho - \rho_0) - c_0^2 \frac{\partial^2}{\partial x_i \partial x_i}(\rho - \rho_0) = -\frac{\partial^2 \phi}{\partial x_i \partial x_j} \qquad\qquad \text{Eq. 2.38}$$

In this context, ϕ is a filtered source term defined to capture only the relevant acoustic sources. It is determined by solving the following Poisson equation:

$$\frac{\partial^2 \phi}{\partial x_i \partial x_j} = -\frac{\partial^2 \rho u_i u_j}{\partial x_i \partial x_j} \qquad\qquad \text{Eq. 2.39}$$

The filtered source term is derived to ensure that the turbulent contributions to the Lighthill stress tensor are appropriately represented without including hydrodynamic influences. Actran's APE sequence incorporates this additional processing step beyond the standard Lighthill approach, filtering out hydrodynamic modes to concentrate on acoustic effects. Implementing the APEs and effectively splitting the acoustic contributions from the hydrodynamic ones, researchers and engineers can achieve more accurate and insightful analyses of aeroacoustic phenomena, leading to better designs and noise mitigation strategies.

3 State of the Art

This chapter reviews the current state of the art in Computational Aeroacoustics for automotive applications and the experimental techniques used for validating numerical models. It provides a comprehensive overview of the latest advancements and methodologies in the field, emphasizing both theoretical and practical aspects.

3.1 Hybrid approaches in Computational Aeroacoustics

This section delves deeper into hybrid approaches within Computational Aeroacoustics. It begins by providing a general overview of CAA methodologies specifically for automotive applications. The discussion then focuses on specific numerical methods based on Lighthill's acoustic analogy and Acoustic Perturbation Equations. Finally, the section concludes with in-depth analysis of the applications and limitations of these hybrid approaches.

3.1.1 Overview on the hybrid CAA approaches

Aeroacoustic problems can be categorized into two types based on the interaction between the flow and acoustics: two-way coupled and one-way coupled scenarios. In a two-way coupled scenario, there is a bidirectional energy exchange where both the flow and acoustics influence each other. Conversely, a one-way coupled scenario occurs when acoustics are affected by the flow, but the flow remains unaffected by the acoustics. In these cases, the problem can be divided into separate flow and acoustics problems. This principle serves as the basis for most aeroacoustic modeling methods.

Theoretically, it is possible to solve simultaneously both the acoustic and flow fields by applying the Navier–Stokes equations over a large domain extending from the source region to the far field. However, this approach encounters significant challenges due to the need to manage fluctuations that span a wide range of length scales and pressure amplitudes. Turbulent structures typically have small length scales but high pressure amplitudes, while the acoustic waves they produce have much longer wavelengths and smaller pressure amplitudes. Consequently, directly computing these interactions demands significant computing resources and is typically feasible only for academic or simplified cases [6].

In practical scenarios, where one-way modeling is feasible, it is essential to divide the problem and address each component individually. Standard automotive applications typically feature flow fields at low Mach numbers, allow for the effective use of one-way coupling. Consequently, the numerical simulation can be divided into four main modeling steps [53, 54]:

1. **Unsteady flow field:** Simulating the transient behavior of flow field variables, including velocity, pressure, and density (if compressibility is considered).
2. **Noise sources**: Identifying and modeling the primary sources of noise.
3. **Exterior acoustic field**: Simulating how noise propagates from the source to a distant observer and interacts with the vehicle's exterior.
4. **Transmission into the vehicle cabin**: Analyzing how noise transmits into the cabin through various car components, such as windows, panels, and seals.

By the use of these methods, widely known in CAA as hybrid approaches, it is possible to develop a comprehensive understanding of aeroacoustic phenomena and devise effective noise reduction strategies while maintaining relatively low computational costs. Additionally, it must be noted that hybrid schemes distinctly separate the flow, aeroacoustic and vibro-acoustic computations. This separation allows for the utilization of an optimal computational grid tailored to each specific physical field, thereby enhancing accuracy [53]. The CFD flow grid is designed to resolve boundary layers with high precision while ensuring sufficient resolution in highly turbulent regions, such as wake areas, to capture essential flow dynamics effectively. It becomes progressively coarser toward outflow boundaries to minimize computational cost while effectively dissipating vortices without compromising the accuracy of the critical flow features. Conversely, the aeroacoustic and vibro-acoustic grids are designed to model the acoustic wave propagation and require a uniform grid size throughout the computational domain.

From a broader perspective, three main aeroacoustic formulations are used to model exterior acoustic fields in automotive applications: Direct Noise Computation (DNC), Acoustic Analogies (AA), and Perturbation Equations (PE).

Schoder provides a comprehensive review of the state-of-the-art computational methods in this field in [53].

DNC, due to the limitations in industrial applications mentioned above, is rarely used in the automotive sector in the framework of Navier-Stokes-based codes. It is typically reserved for specific scenario, such as modeling side mirror feedback phenomena as shown in [55], where the interaction between acoustic and flow fields is crucial. In contrast, Lattice Boltzmann-based CFD solvers, which are inherently transient and compressible, are employed in a considerably wider range of applications due to their more affordable computational costs. [56].

AA transform complex fluid dynamics equations into more tractable forms, facilitating noise prediction and analysis. Several notable acoustic analogies have been developed over the years, each with its unique approach and application scope. Lighthill's pioneering work [11] transformed the compressible Navier-Stokes equation into an exact inhomogeneous wave equation, known as Lighthill's equation (see section 2.2.3). Curle [57] extended Lighthill's theory by investigating the effects of surfaces at rest, equating them to surface dipole distributions. Ffowcs Williams and Hawkings (FW-H) [58] further extended Kirchhoff's formula [59], generalizing the integral solution to account for arbitrary moving bodies within the source domain. These advancements allowed for more accurate modeling of noise in the presence of solid boundaries and moving surfaces. Ribner [60] and Powell [61] developed theories based on the comparison of slightly compressible flow with incompressible flow, where vorticity sources are dominant. Their work, along with further investigations by Howe [46, 47, 62], Möhring [63], and Doak [64], contributed significantly to the understanding of vortex sound. These studies focused on the noise generated by vortices, which are common in various aerodynamic applications. Goldstein proposed a generalized acoustic analogy [49], demonstrating that the Navier-Stokes equations could be reformulated as a set of Linearized Euler Equations (LEE). This approach provided a broader framework for analyzing aeroacoustic problems, accommodating a wider range of flow conditions and source characteristics.

PE are based on the systematic decomposition of the field properties into fluid dynamic and acoustic components. Unlike traditional aeroacoustic analogies, this method allows the acoustic component to be analyzed within the flow domain, including the near field. The linearized Euler equations (LEE) are derived by decomposing field variables such as density, velocity, and pressure into their mean and fluctuating parts. Over the years, modifications to the LEE have been made to ensure only acoustic waves are propagated. Notable contributions include Ewert and Schröder's filtering techniques [13], Ribner's dilatation equation [60], and Hardin and Pope's Expansion about incompressible flow method [65]. While these methods have been validated against various test cases, they have also faced criticism [66] and subsequent modifications to account for non-isentropic flows and near-field coupling effects. More recent advancements, such as the perturbed compressible equations (PCE) and Acoustic Perturbation Equations (APE) (see section 2.2.4), offer improved modeling capabilities for low Mach number flows. Researchers like Munz et al. [67, 68] have further refined these approaches by incorporating Mach number scaling into the compressible Euler equations. Additionally, techniques based on Helmholtz decomposition [69] have been proposed to separate aerodynamic and acoustic components more effectively.

In addition to the aeroacoustic methods discussed, vibroacoustic modeling is crucial for understanding and mitigating interior noise transmission within a vehicle cabin. This involves analyzing how structural vibrations, induced by exterior acoustic fields, interact and propagate through vehicle components and convert into perceived noise inside the cabin. Numerous VA approaches have been proposed [70, 71, 72, 73]. Among these, Boundary Element Method (BEM) [74] and Finite Element Method (FEM) [75], which is used in this work, are widely employed to model the propagation of hydrodynamic and acoustic fluctuations through different media and their interaction with vehicle structures, such as windows and panels. BEM is often used for its efficiency in handling unbounded acoustic domains, while FEM is preferred for its versatility in dealing with complex geometries and material properties. For high-frequency noise prediction Statistical Energy Analysis (SEA) is applied, where traditional deterministic methods become computationally expensive. SEA divides the vehicle structure into subsystems and uses statistical methods to predict energy distribution and sound transmission [76]. Furthermore, advanced modeling techniques can include the reciprocal fluid structure interaction, which simulate both the deformation of the structure caused by the fluid and the subsequent impact of this deformation on the fluid behavior [1].

By combining aeroacoustic simulations and vibroacoustic analysis, it is possible to develop comprehensive noise reduction strategies that address both the generation and transmission of noise, ensuring a quieter and more comfortable vehicle interior. These methodologies, validated through experimental data and real-world testing, enhance the efficiency and precision vehicles design, ensuring compliance with stringent acoustic comfort standards.

3.1.2 Application of hybrid CAA approaches in the automotive industry

Among the computational approaches discussed, those based on acoustic analogies have found significant success and are widely used in automotive applications, as evidenced by numerous references. These methodologies have been extensively explored in various studies, demonstrating their versatility and effectiveness in addressing complex aeroacoustic problems.

One prominent area of application is the investigation of side-view mirrors geometries. For instance, numerous researchers have employed hybrid approaches to predict and analyze the noise generated by side mirrors. The findings from these studies have provided valuable insights into numerical methodologies, highlighting both their limitations and strengths. They have also demonstrated the effectiveness of computational processes in optimizing the design of quieter and more aerodynamically efficient mirrors. For instance, Lokhande [77] demonstrated a hybrid approach using LES and FW-H acoustic analogy on a simplified mirror shape proposed by Höld et al. [78] and Siegert at al. [79]. The SPL at a far-field noise receiver closely matched experimental measurements up to 1 kHz, with a mean deviation of less than 5dB. Similarly, Khalighi [80] tested various CFD simulations, including URANS, DES, and LES, coupled with Curle's formulation to predict pressure fluctuations. The study showed reasonable agreement between virtual surface microphones and experimental results, particularly with the LES approach, which more accurately predicted surface pressure distributions and sound levels. The simulations indicated that early flow separation, influenced by mirror shapes, significantly impacts the noise generated. Chode et al. [81] presented alternative approaches employing a hybrid RANS/LES coupled with FW-H and APE to predict near and far field noise. This study showed convincing correlation up to 2 kHz and identified the main geometric features causing major noise. Chen et al. [82] conducted a similar study by coupling LES with FW-H, measuring unsteady pressures and noise levels at different flow speeds and for two different side mirror configurations. Excellent correlations with experimental results were observed up to 5 kHz at surface microphones installed downstream of the side mirror wake, demonstrating the beneficial impact of controlling the flow in the recirculation zone on noise reduction.

Similarly, the hybrid approaches based on AA and VA methods have been applied to the analysis of simplified vehicle models such as SAE body and Hyundai Simplified Model (HSM). These studies focus on the noise generated by simplified vehicle geometries, providing insights into the fundamental aeroacoustic phenomena including transmission to the interior. By analyzing these simplified models, researchers primarily validate and compare numerical tools, ensuring their reliability and accuracy. Additionally, these studies have helped clarify certain noise generation mechanisms, providing insights that contribute to the development of noise reduction strategies in real-world applications. In this context, Hartmann et al. [37] conducted a comprehensive study on the noise generated by airflow around the A-pillar and side mirror of an SAE model. They simulated pressure fluctuations on the window under different flow conditions and vehicle setups, validating their results against wind tunnel experiments. Using various CFD solvers and turbulence models (including incompressible DES based on SA), they achieved accurate predictions up to 2 kHz, with increasing deviations beyond this frequency. The study found that acoustic excitations were the main contributors to interior SPL, rather than hydrodynamic excitations. Similarly, Jacqmot et al. [83], achieved convincing results in modeling both the pressure fluctuations on the side window and the interior SPL up to 6 kHz using compressible CFD combined with FEM. By employing the APE in the process, the study highlighted the significant role of acoustic contributions to interior noise, consistent with previous findings. They also noted that pressure waves with intermediate wavelengths between hydrodynamic wavelengths and acoustic wavelengths can interact with the window at frequencies below the coincidence frequency. Blanchet et al. [73] demonstrated that combining a compressible DES with a VA model using a hybrid FE/SEA approach leads to accurate interior SPL predictions up to 7 kHz. Their model was proven to be robust and reliable in simulating various design changes. Ganty et al. [10] proposed an aero-vibro-acoustic simulation for the HSM by coupling a Lattice-Boltzmann-based CFD code with a vibro-acoustic code based on FEM in Actran. This approach accurately predicted A-pillar wind noise behavior up to 4.5 kHz, making it suitable for industrial vehicle development. A relevant study in this context was carried out by Cho et al. [84], who conducted a comprehensive benchmark combining various CFD numerical methods (including FVM and LBM), turbulence models (such as LES, DDES, and IDDES), and VA approaches (SEA and FEM). Their simulated interior SPL demonstrated high accuracy across all proposed approaches up to 4 kHz. Similar studies conducted on SAE body and HSM can be found in references [85, 86, 87].

Complete vehicle simulations have also benefited from hybrid approaches. Full-scale vehicle models pose a greater challenge due to their complexity and significant computational cost. However, studies employing AA have demonstrated significant progress in predicting and mitigating noise in these scenarios. Simulations applied to production vehicles help in understanding the overall noise generation and propagation mechanisms, ultimately aiding in the development of quieter vehicles. Among the notable studies in this field, Pietrzyk et al. [7] applied a hybrid approach to assess exterior noise generated by a production vehicle at 140 km/h, focusing on the side mirror and A-pillar regions. By combining Large Eddy Simulation (LES) with the Smagorinsky turbulence model and a Finite Element (FE) model using Lighthill's analogy in Actran, they achieved strong correlations with wind tunnel tests for propagated SPL up to 2 kHz. The study also underscores the influence of CFD grid size on the cut-off frequency, that determines the maximum frequency that can be modeled. Yuan at al. [88] combined IDDES and APE to study hydrodynamic and acoustic pressure fluctuations on the side window of a commercial road vehicle. The simulations, validated with surface microphones, showed good agreement up to 2 kHz. Beyond this frequency, the DES model underestimated the hydrodynamic contribution, highlighting limitations in wall modeling. The results revealed that hydrodynamic pressure is concentrated on the front side window, especially in areas influenced by the side mirror wake, while acoustic pressure is uniformly distributed with energy spread across a wide frequency range. Similarly, Zhong et al. [89] proposed a hybrid CAA method in conjunction with wind tunnel experiments to investigate the aerodynamic noise generated from a passenger vehicle. Specifically, a DES/APE approach was used to model the exterior noise while an SEA approach was used to predict the interior noise transmission. Results showed good correlations against experiments up to 5 kHz. The study also provided a comprehensive analysis of the primary sources affecting interior SPL. The findings indicated that noise impacting front passengers is predominantly originated from the side windows, while rear passengers are mainly affected by noise from the rear side windows and the rear window.

Overall, these studies underscore the high accuracy achievable by hybrid computational approaches under various flow conditions and model complexities across a broad frequency range. Hybrid RANS/LES simulations have proven effective in capturing both narrow-band and broadband fluctuations. When properly coupled with Aero-Vibro-Acoustic (AVA) models, they can deliver highly accurate results for both exterior and interior noise. Research indicates the capability of characterizing primary noise sources and correlating them with transient turbulent structures through combined flow and acoustic field analyses. Numerous studies have identified the acoustic contribution as the dominant factor in interior noise transmission mechanisms. Additionally, these studies emphasize that the selection of the most appropriate computational process in vehicle development is influenced by factors beyond accuracy. These include pre-existing knowledge, versatility across a wide range of applications, and available resources.

Despite extensive research and numerous studies employing hybrid approaches, a significant gap remains in the literature. Specifically, there is a lack of comprehensive studies that fully document all steps of the validation process. Due to the confidentiality constraints within the industrial framework, public and open-source research involving full production vehicles is scarce, as noted previously. This leads to a shortage of documentation on creating and validating numerical simulation processes from scratch and integrating them into vehicle development scenarios. Furthermore, the literature inadequately addresses the integration of validation methodologies with experimental data, not only for validation purposes but also within the vehicle development framework. This integration is crucial for understanding the benefits and limitations of these methods in industrial applications. Additionally, most studies focus on side mirrors and the A-pillar region, with no comprehensive analysis of other region or component behavior such as a roof spoiler on road vehicles. This further underscores the need for a more detailed and integrated approach to vehicle aeroacoustic analysis.

3.1.3 Limitations of the hybrid CAA approaches

Hybrid CAA approaches, which combine CFD and AVA models in series, are subject to inherent limitations at each modeling phase.

Firstly, all the limitations of hybrid RANS/LES modeling (see chapter 2.1.6 in the context of DDES) must be taken in consideration. For instance, the use of wall models may not ensure high accuracy in regions with highly turbulent boundary layers or massive separation areas. Predicting hydrodynamic behavior using these models can result in significant deviations, especially at high frequencies, as demonstrated in [88]. Additionally, the solution's accuracy is heavily dependent on the chosen turbulence model and the grid used [7, 89]. While coarsening the grid can reduce computational costs and create the impression of a more efficient process, it may lead to inaccurate modeling and erroneous results.

Furthermore, limitations related to acoustic analogies must be acknowledged. These methods operate on the principle that noise generation and propagation are independent, meaning the generated noise does not influence the flow field [8]. Accurate modeling of the interaction between the acoustic field and the flow field requires the use of compressible flow equations, moving beyond the assumptions of acoustic analogies, to resolve all turbulence and wave structures. However, in hybrid aeroacoustic methods, including the approach used in this work, incompressible flow equations are often applied, ignoring the effect of the acoustic field on the flow field. Consequently, specific phenomena, such as side mirror feedback noise [55] can only be predicted using compressible simulations [53]. Thus, hybrid CAA approaches are inherently unidirectional. For instance, the approach based on Lighthill's acoustic analogy, as employed in this work, is specifically tailored for low Mach number scenarios where convection and refraction effects during propagation are negligible for the geometric scales with vehicles [52]. This implies that the Lighthill's analogy is suitable for flows with Mach numbers below 0.2, where these effects are relatively minor. For higher Mach number flows, alternative analogies, such as Möhring's analogy, are recommended to account for significant convection and refraction effects.

Finally, also applications of vibro-acoustic models based on FEM or SEA present several limitations that must be noted [1, 54, 70]. The FE-based VA models, while highly accurate in capturing detailed structural responses, suffer from significant computational demands, especially when applied to high-frequency ranges or large-scale problems. This constraint often limits their practical use in real-time or complex simulations. On the other hand, SEA-based VA models, though more computationally efficient and suitable for high-frequency analysis, rely heavily on statistical assumptions and averaged properties, which can lead to less precise predictions for systems with intricate geometric or material variations. Moreover, SEA models are less effective in low-frequency regimes where deterministic approaches like FE are more appropriate.

Despite the demonstrated effectiveness of hybrid CAA approaches in the automotive industry (see section 3.1.2), these limitations underscore the need for careful application, especially in industrial contexts. Accurate modeling is critical in vehicle development and has a significant impact on decision-making and costs. Therefore, understanding and mitigating these limitations is essential for the successful implementation of hybrid CAA methods in aeroacoustic development processes.

3.2 Experimental approaches for validation of aeroacoustic simulations

In this section experimental methods for validating aeroacoustic models are examined, covering the use of aeroacoustic wind tunnels and techniques for measuring and analyzing flow fields, exterior noise, and interior noise.

3.2.1 Aeroacoustic wind tunnels for vehicle development

Aeroacoustic numerical methods employed in the automotive industry necessitate rigorous validation to ensure accuracy and reliability, which are considered crucial aspects in the development process. This validation is typically achieved through experiments that offer real-world data effectively in a reproducible way, allowing for multiple measurements and configurations within a relatively short timeframe. In contrast, simulations would struggle to reproduce the same number of configurations within such time constraints. Wind tunnels play a crucial role in this context, serving as specialized facilities designed to reproduce and analyze the aerodynamic and acoustic performance of vehicles under controlled conditions. Unlike on-road tests, wind tunnels ensure a high degree of reproducibility, which is generally very challenging to achieve in real-world settings [1, 19].

There are two main types of wind tunnel designs based on the type of air return: Göttingen and Eiffel. The Göttingen wind tunnel features a closed-circuit design where air is driven by a fan, circulating continuously within the tunnel. This design allows for precise control of airflow conditions, ensuring consistent and repeatable testing environments. Additionally, it is efficient as the flow energy can be reused, although it comes with the drawback of significantly higher construction costs. On the other hand, the Eiffel wind tunnel operates on an open-circuit principle, where air is drawn from the environment and expelled back out after passing over the test model. This design is often simpler and can be more cost-effective, though it may be influenced by external atmospheric conditions.

In the automotive sector, wind tunnels are configured to replicate a variety of conditions to facilitate comprehensive aerodynamic and aeroacoustic vehicle development. These configurations can include varying wind speeds, yaw angles, and turbulence levels to reproduce different realistic driving scenarios. Advanced measurement techniques, discussed in sections 3.2.2 and 3.2.3, are typically employed to capture detailed aerodynamic and acoustic data. These measurements help optimizing aerodynamic performance, in identifying and mitigating noise sources, and hence improving overall aeroacoustic vehicle design.

Different wind tunnels are designed to serve various specific purposes in vehicle development [1]. Full-scale wind tunnels, for instance, are capable of testing entire vehicles at 1:1 scale, providing highly accurate data on aerodynamics performance. Aeroacoustic wind tunnels, a subset of full-scale wind tunnels, are specially designed to study the noise generated by vehicles. In such facilities, careful control of background noise is essential, as various sources can affect the acoustic measurements in the test section. These include noise from the drive unit, airflow through the intake and fan, as well as flow noise generated by the airline, its components, and the test section itself, all of which depend on the specific design of the wind tunnel. To mitigate these issues, fan noise must be isolated from the test section. Components such as safety grids, heat exchangers, flow straighteners (e.g., honeycombs), and turbulence-generating grids must be carefully positioned and designed to ensure their noise remains undetectable in the test area, maintaining the accuracy of acoustic measurements. Moreover, the test section and other key components, such as the first diffuser, turning vanes, and splitter walls, are typically equipped with sound-absorbing linings to minimize noise interference. These tunnels are equipped with advanced acoustic measurement systems, including large microphone array systems and sound pressure sensors, to pinpoint and analyze noise sources [19, 43].

Aeroacoustic wind tunnels offer automotive engineers a versatile and comprehensive toolset for vehicle development. Their holistic design ensures that the data collected is reliable and closely aligned with real-world driving conditions, contributing to the creation of better-performing and more efficient vehicles.

3.2.2 Flow field measurements and analysis techniques

Flow field measurements and analysis are critical in aerodynamic and acoustic testing, particularly in the context of vehicle development within wind tunnel environments. These measurements are essential for understanding how air moves around a vehicle, influencing both its aerodynamic efficiency and noise characteristics. Various techniques and instruments have been developed and refined to capture this data with high precision [1]. In the present work, we utilized static pressure probes, pressure rakes, and Cobra probes, all of which are described in this section

Static pressure probes are designed to measure the static pressure on a vehicle's surface. These probes are typically thin, disc-shaped devices with a connecting tube, often welded to the probe and positioned radially towards the center, where the tube features a hole of 0.4 mm or smaller (see **Figure 3.1** left). For optimal integration, the probes can be flush-mounted on the vehicle's surface using adhesive foil, ensuring stability and minimal flow disturbance, allowing a large number of measurement points to be recorded during testing. The accuracy of these measurements is highly sensitive to the size and quality of the pressure hole, which must be less than 1 mm in diameter, with edges that are either rounded or sharply defined, and free from any burrs to prevent deviations [90]. These probes enable detailed understanding of the pressure development along the vehicle's surface, providing valuable data for aerodynamic analysis. Additionally, in the context of CFD modeling, they are exploited to validate wall modeling.

Pressure rakes, typically consisting of an array of Pitot tubes (see **Figure 3.1** center) arranged in a grid, are used to perform volume measurements and acquire dynamic pressure data. These devices measure the total pressure (p_{tot}) within a flow, which, when combined with static pressure (p_{stat}), enables the calculation of dynamic pressure (p_{dyna}). The configuration of the Pitot tubes in a pressure rake allows for pressure measurement across a section of the flow field, providing valuable data for aerodynamic analysis. However, their accuracy can be affected in highly turbulent regions, particularly at high angles. The shape of the probe head plays a crucial role in measurement accuracy, and precise alignment is essential to ensure reliable results [91]. In the context of numerical model validation, they are also used to validate the spatial pressure distribution, which is often linked to the grid refinement strategy and turbulence model employed.

Cobra probes, featuring four holes (see **Figure 3.1** right) capable of measuring four pressure differences, are highly accurate instruments with a measurement precision of ±0.3 m/s and an angular accuracy of ±1°. They can sample data at frequencies up to 1550 Hz [1], making them well-suited for capturing dynamic flow characteristics. Due to their sensitivity, misalignment can lead to significant errors, necessitating extensive calibration time in the wind tunnel. Unlike traditional single-component measurement devices, Cobra probes are designed to capture three-dimensional, timely resolved velocity components, providing a comprehensive understanding of both the speed and direction of airflow as well as the turbulence. This capability enables detailed mapping of complex flow structures and the acquisition of turbulent data, allowing engineers to conduct both transient and time-averaged analyses. Recent advancements in Cobra probe technology have focused on enhancing spatial resolution and reducing the physical size of the probes to minimize their impact on the flow. These improvements enable researchers to capture detailed and accurate data on flow phenomena that were previously challenging to measure.

The measurement devices discussed above provide critical data that enhance our understanding of airflow behavior around vehicles. As technology progresses, the precision and reliability of these measurement techniques will continue to evolve, offering even greater insights into aerodynamic and aeroacoustic phenomena.

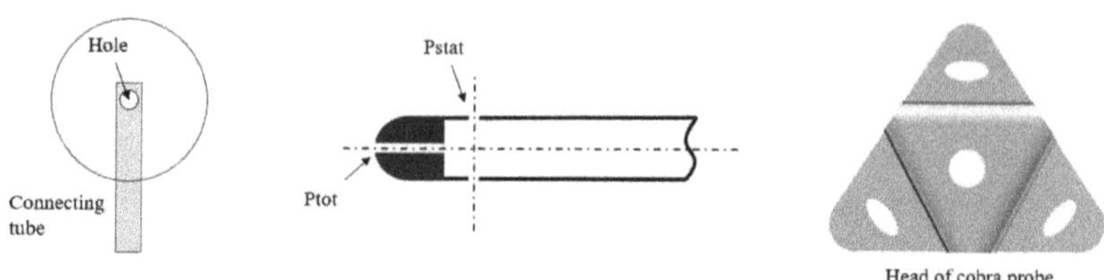

Figure 3.1:   Schematic representation of a flat pressure probe (left), a pitot tube (center) and a generic 4-holes head of a cobra probe (right).

3.2.3   Aeroacoustic measurements and analysis techniques

In the context of wind tunnel testing, acoustic measurements are essential not only for vehicle development but also for validating numerical models and ensuring that simulation results accurately reflect real-world conditions. This work includes several aeroacoustic measurements taken with the following devices: exterior and interior microphones, surface microphones, and microphone arrays.

Exterior microphones are commonly used to measure noise sources emanating from specific regions of a vehicle [1]. However, the challenge lies in minimizing the hydrodynamic contribution to the measurements. In order to address this criticality, the microphone can be installed on an aerodynamic arm equipped with nose cones designed to reduce flow disturbance (see **Figure 3.2** left). This setup ensures that the microphones are positioned close to the noise sources while remaining outside the turbulent flow, thereby capturing more accurate data. Despite these efforts, a critical issue remains: the interaction of the microphone arm and tip with the flow can still introduce unwanted noise, complicating the comparison with simulation data where such disturbances are typically not accounted for.

In addition to individual exterior microphones, microphone arrays (see **Figure 3.2** center) are employed to perform beamforming, a technique that allows for the spatial localization of noise sources [92, 93]. These arrays are typically installed outside the shear layer of the wind tunnel, which is essential to avoid interference from turbulent flows but introduces additional complexities when comparing measurements with simulations. One of the primary challenges is accounting for refraction and convection phenomena, which can significantly affect the accuracy of beamforming results. Refraction, caused by varying flow velocities within the wind tunnel, can bend the acoustic waves, leading to an apparent shift in the perceived location of the noise source. Convection, here the transport of the sound waves by the moving airflow, can alter the speed and direction of the sound, further complicating the interpretation of the data. Moreover, beamforming techniques rely heavily on the coherence of the signal, which can be compromised by the complex flow environment around the vehicle, especially at higher frequencies. These effects are difficult to replicate in numerical models, which often assume a uniform flow field and may not fully capture the complex interactions present in a real wind tunnel environment. This disparity between the measured and simulated data poses a significant challenge in validating aeroacoustic simulations.

Surface microphones (see **Figure 3.2** right), placed directly on the vehicle's body (e.g., on the car's windows), provide valuable data on the pressure fluctuations caused by aerodynamic forces. These microphones are particularly effective in capturing the noise contribution from the boundary layer and other near-surface phenomena. However, surface microphones inherently capture both hydrodynamic and acoustic contributions, with their response influenced by the microphone size and the wavelength or scale of the flow. Their measurements can be further contaminated by additional turbulence generated by the presence of the microphones themselves. Moreover, the measurements are highly sensitive to the exact placement on the vehicle's surface, which can introduce discrepancies when validating simulations, particularly if the boundary layer characteristics are not perfectly replicated in the numerical model.

Interior microphones, including artificial heads, are used to measure the noise transmitted into the vehicle's cabin, which is critical for assessing passenger comfort [43]. These devices capture the contribution of exterior sources to the interior noise environment. A significant challenge in simulations involving interior microphones is accurately replicating the real-world conditions within the vehicle. The complex interactions between the vehicle structure, interior materials, and exterior noise sources can be difficult to model numerically, making it challenging to ensure the simulated data reflects the actual acoustic environment. The artificial head, designed to mimic the acoustic properties of a vehicle's passenger, helps in approximating the perceived noise levels, but it introduces further complexity in the validation process due to the necessity of accurately modeling the acoustic properties of the human ear.

In conclusion, while aeroacoustic measurements using various microphone configurations provide accurate and effective data for vehicle development, they also introduce significant challenges in terms of comparing this data with simulations. Each type of microphone setup has its own criticalities, particularly in terms of flow disturbances, coherence of the signal, refraction and convection effects, and sensitivity to placement, all of which must be carefully considered to ensure accurate validation of numerical models.

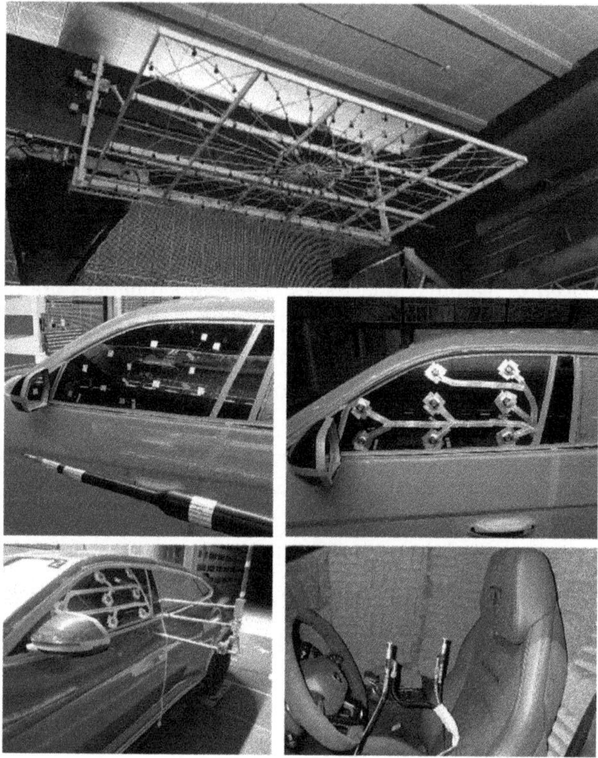

Figure 3.2: Measurement devices: microphones array (top), exterior microphone with nose cone and surface microphones (center), cobra probes and interior microphones (bottom)

4 Numerical Approach

In this chapter, the developed numerical process is presented. The methodology and the computational steps involved are comprehensively discussed to provide insight into the design of the simulation. Specifically, the hybrid numerical approach is presented for two specific areas: the side mirror and the roof spoiler region.

4.1 Hybrid computational aeroacoustic process

The vehicle analyzed in this investigation is the SUV Lamborghini URUS (**Figure 4.1**). All simulations were conducted under steady-state operative conditions with the vehicle positioned at 0° yaw and traveling at a speed of 140 kph. The focus of the investigation was directed toward the side mirror and roof spoiler components, along with their surrounding regions, which were thoroughly analyzed to evaluate their aeroacoustic behavior.

Figure 4.1:   Reference vehicle model: Lamborghini Urus

The numerical process investigated in this work consists of a three-step hybrid approach resulting from combining a CFD and two FE models. In the first step, the transient velocity field in the investigated region (i.e., side mirror and roof spoiler) is solved and extracted by an incompressible CFD simulation

© The Author(s), under exclusive license to
Springer Fachmedien Wiesbaden GmbH, part of Springer Nature 2026
C. A. Perugini, *An Efficient Hybrid Computational Process for Aeroacoustic
Vehicle Development Based on Navier-Stokes Equations and Lighthill's
Acoustic Analogy*, Wissenschaftliche Reihe Fahrzeugtechnik Universität
Stuttgart, https://doi.org/10.1007/978-3-658-51977-3_4

performed with OpenFOAM. Subsequently, the acoustic field is calculated in the same region by means of a finite element analysis based on the Lighthill's analogy and the acoustic perturbation equations performed by Actran. Both the total and the acoustic pressure fluctuations are then calculated on the car surface as well as on the specific window under investigation (e.g., the side window in the case of the side mirror, and the rear window in the case of the roof spoiler). Based on this result, a vibro-acoustic model is used to simulate the transmission of the aeroacoustic pressure fluctuation and the consequent SPL propagation into the cabin. In this model, the structural transmission through the glass is considered as the boundary condition. **Figure 4.2** provides an overview of the primary steps in the computational aeroacoustics workflow, using the side mirror region as a representative example.

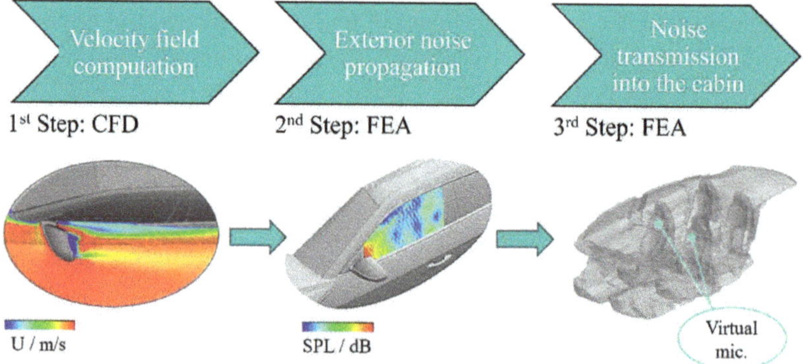

Figure 4.2:   Workflow of the CAA process applied to the side mirror region: computational steps and main outcome

It is important to emphasize that the process is designed for localized analysis of the vehicle. Consequently, the procedure is applied separately to each specific area under investigation: side mirror and roof spoiler. A detailed description of the above-mentioned computational steps is provided in the following sections.

4.2 DDES-based unsteady velocity field calculation

The simulations were performed within a simplified computational domain (see **Figure 4.3**), applying specific boundary conditions tailored to replicate the experimental aerodynamic and aeroacoustic behavior while minimizing boundary interference effects. The freestream velocity was imposed at the inlet, and the atmospheric pressure condition was assigned at the outlet. Inviscid flow conditions were applied to the lateral walls and roof to reduce wall-induced disturbances. The floor boundary was segmented into two regions: from the inlet to the front of the vehicle, an inviscid condition was implemented, while a no-slip boundary was imposed from the vehicle to the outlet. This configuration supports accurate boundary layer development across the vehicle's surface, enhancing the simulation's fidelity to physical flow behavior. The computational domain dimensions and boundary conditions were established based on a prior study that optimized the model to align with wind tunnel experimental data (see Chapter 5). To maintain focus, the specifics of this preliminary study are not included in this work.

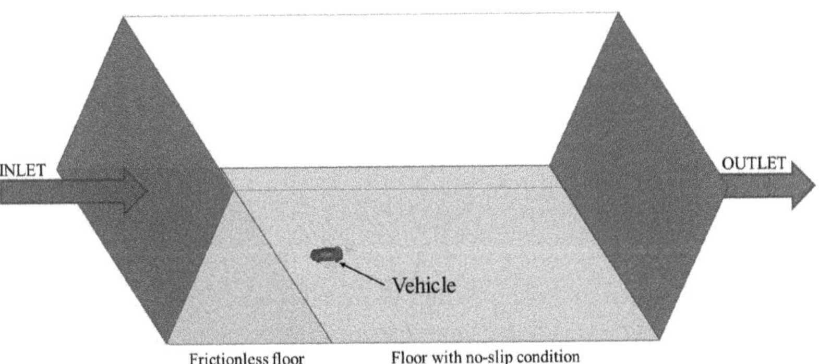

Figure 4.3:   Simulation domain dived into regions to illustrate the application of distinct boundary conditions

Under the assumption of incompressible flow, a delayed detached eddy simulation (DDES) with the Spalart-Allmaras turbulence model (see Section 2.1.5) was performed to calculate the transient velocity field around the entire vehicle. The CFD simulation was initially run for 2 seconds of physical time in order to reach the convergence of the residuals and the stability of the main fluid dynamic quantities. Subsequently, velocity vector data within an extraction volume corresponding to the specific region under investigation (i.e., side mirror or roof spoiler, as shown in **Figure 4.4**) were recorded over a time span of 0.2 seconds.

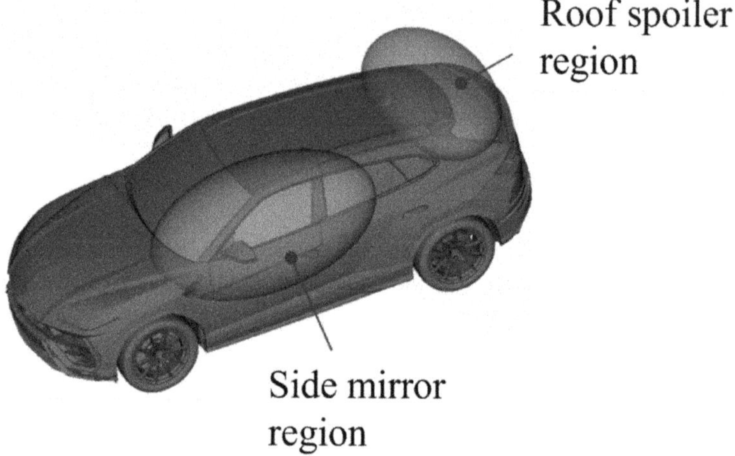

Figure 4.4:   CFD model with high-refined volumes used to solve and extract the velocity field from the side mirror and the roof spoiler regions

The simulations were conducted considering a free stream velocity (U_∞) of 140 km/h. A structured grid of 2 mm was used to discretize the velocity extraction volume, representing the most refined region of the mesh domain. To satisfy the Courant-Friedrichs-Lewy (CFL) condition (see Eq. 4.1), the simulation employed a time step of $2.5 \cdot 10^{-5}$ s.

$$\frac{U_{max}\Delta t}{\Delta_{min}} < 1 \qquad\qquad \text{Eq. 4.1}$$

Here, U_{max} is the maximum expected velocity, Δt the simulation timestep and Δ_{min} the smallest cell size in the CFD grid (corresponding to the cell size in the velocity extraction volume). Based on a maximum expected velocity of 1.5

times the free stream velocity and a smallest cell size of 2 mm, the required time step was calculated using Eq. 4.1.

Furthermore, a High-Reynold approach by means of wall functions was employed to model the boundary layer. As discussed in Section 2.1.4, this approach requires a dimensionless wall distance (y^+) greater than 30. Using the friction velocity and the kinematic viscosity as characteristics of the investigated fluid (i.e., air), the correct range of y^+ was achieved by setting the first prism layer thickness (d_{PL}) to 0.5 mm. By geometric definition, d_{PL} is twice the distance of the grid centroid to the nearest wall (d).

After configuring the simulation parameters, the sampling frequency (f_s) for extracting the unsteady velocity field over 0.2 s of physical time was determined. The Nyquist-Shannon theorem [94] (Eq. 4.2) was applied to ensure the sampling frequency was sufficient for accurate aeroacoustic analysis:

$$f_{max} \leq \frac{f_s}{2}$$

Eq. 4.2

The Nyquist-Shannon theorem provides a sufficient condition to sample correctly a time signal of a finite bandwidth. Targeting a maximum frequency (f_{max}) of 5 kHz for the simulation, a sampling frequency of 10 kHz was selected, corresponding to an extraction timestep (Δt_{ext}) of 10^{-4} s.

Figure 4.5 provides a schematic summary of the CFD simulation steps as a function of physical time.

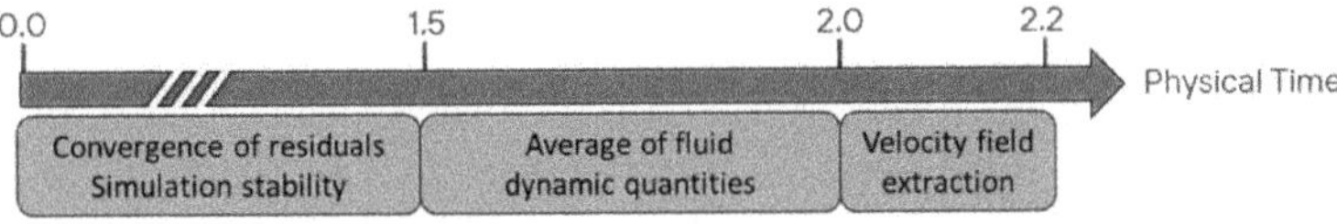

Figure 4.5:   CFD computational phases as a function of physical time

The first 1.5 s of the simulation were necessary to achieve the convergence of the residuals and the stability of the main fluid dynamic quantities. The next 0.5 s (from 1.5 s to 2 s) were used to average over time the main flow quantities for validating of the CFD model. This time period also provided a buffer between the start of the velocity extraction process and the initial convergence phase of the simulation. This approach ensured the reliability and accuracy of the extracted velocity data and avoided distortions caused by transient effects during the early stages of the simulation. Lastly, velocity vectors are extracted within an elliptical extraction volume (see **Figure 4.4**) over a period of 0.2 s, using the sampling frequency calculated previously.

It is observed that the simulation setup presented is valid for both the side mirror and roof spoiler regions. An overview of the main parameters required for the simulation setup is provided in the following table.

Table 4.1:   CFD simulation setup and main parameters

Smallest cell size (Δ_{min}) / mm	2
Simulation timestep (Δt) / s	$2 \cdot 10^{-5}$
Velocity extraction timestep (Δt_{ext}) / s	10^{-4}
Sampling frequency (f_s) / kHz	10
First prism layer thickness (d_{PL}) / mm	0.5
y^+ / -	> 30
Physical time for simulation convergence / s, (range)	2, (0.0 s - 2.0 s)
Physical time for average of the fluid dynamic quantities / s, (range)	0.5, (1.5 s - 2.0 s)
Physical time for velocity field extraction / s, (range)	0.2, (2.0 s - 2.2 s)

4.3   Aeroacoustic source modeling and noise propagation

Distinct acoustic domains were defined for the side mirror and the roof spoiler respectively, as shown in **Figure 4.6**. Each FE model was divided into two key regions: the source region (highlighted in green), where the acoustic sources were calculated, and the buffer region (highlighted in orange), used to ensure the convergence of the numerical model and to deploy a non-reflecting boundary condition to the fluid boundary. In this latter region, no sources were calculated. A reflecting boundary condition was assigned to the car surface. To enhance the accuracy of the model, a weighting function was applied near the boundary of the source region to gradually attenuate the computed sources. This spatial filtering technique prevents the generation of spurious noise sources caused by truncation effects, which may occur as vortices cross the border of the source region. **Figure 4.6** provides a detailed depiction of the acoustic domains employed for the side mirror (left) and roof spoiler (right) regions, highlighting the structure of the model.

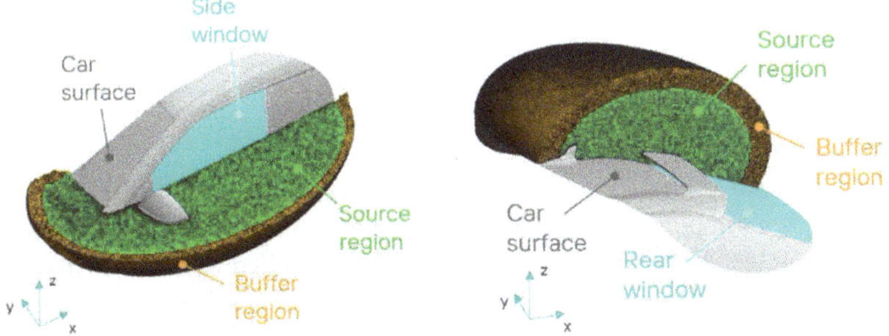

Figure 4.6:   FE models: acoustic domains for the side mirror (left) and the roof spoiler case (right)

The acoustic domain was discretized using a FE mesh, designed to fulfil the requirement of a minimum of 4 quadratic elements per smallest acoustic wavelength (λ_{min}). For a target maximum frequency (f_{max}) of 5 kHz, a maximum permissible element size (L_{max}) of 17 mm was determined using the following equations:

$$\lambda_{min} = \frac{c}{f_{max}}$$

Eq. 4.3

$$L_{max} = \frac{\lambda_{min}}{4} = \frac{c}{4\,f_{max}}$$

Eq. 4.4

The acoustic sources were calculated by Actran performing the FE formulation of Lighthill's analogy, under the assumption of neglecting the entropy and the viscous effects (see section 2.2.3). Based on the velocity field solved and stored in the first step of the process, the acoustic analogy allows to calculate the source term (RHS of Eq. 2.31) as a function of time, encompassing all the type of acoustic sources (i.e., monopoles, dipoles and quadrupoles).

After the computation of the acoustic sources, a Discrete Fourier Transform (DFT) was performed in order to convert the time-domain data into the frequency domain. With a target frequency resolution (Δf) of 20 Hz, a DFT window length (T_{DFT}) of 0.05 s was calculated by means of the following formula:

$$\Delta f = \frac{f_s}{N} = \frac{1}{T_{DFT}} \qquad\qquad \text{Eq. 4.5}$$

where f_s is a sampling frequency of 10 kHz (see **Table 4.1**) and N is the number of samples. A Hanning window function was applied with a 50% overlap between consecutive windows, resulting in a total of seven windows for the DFT, as shown in **Figure 4.7**

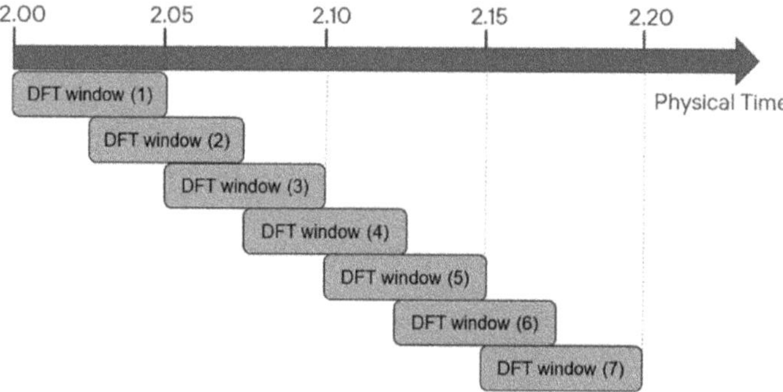

Figure 4.7:   DFT windows distributed over the physical time

Before calculating the exterior propagation of noise, it may be beneficial for aeroacoustic analysis to isolate the acoustic contribution from the total pressure fluctuations, excluding hydrodynamic effects. This was achieved by applying the Acoustic Perturbation Equations (APEs) to the acoustic sources already calculated in the frequency domain. Two types of calculations for exterior propagation are possible:

- **Total pressure fluctuation (PFL)**: Calculated directly over the vehicle surface based on the unfiltered source term in the frequency domain.
- **Sound Pressure Level (SPL)**: Propagated over the vehicle surface and surrounding space using the source term filtered through the APEs.

A comparison of these two calculations for the same window is illustrated in **Figure 4.8**, using the side mirror region as a reference. The left side of the figure shows the PFL distribution, while the right side displays the SPL distribution. While the color scales are shown, the actual range values are omitted due to confidentiality constraints. It is important to note that different range values are used for the two distributions, as the PFL (right-hand side) reaches a maximum value approximately 40 dB higher than the correspondent SPL (left-hand side. This substantial difference arises from the fact that the PFL map includes both acoustic and hydrodynamic pressure fluctuations, whereas the SPL map includes only the acoustic component.

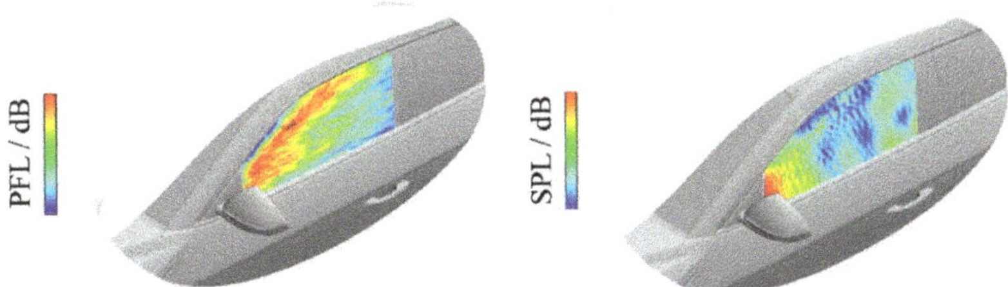

Figure 4.8:　　Comparison between the PFL and the SPL distribution on the side window at 1 kHz (3rd octave band visualization)

The computational steps performed by the FE model in this second stage of the CAA workflow can be summarized as follows:

1　　**Aeroacoustic Source Calculation:** Potential acoustic sources in the time domain were calculated in the source region using Lighthill's analogy, based on the velocity field modeled in the CFD stage.

2　　**Frequency Domain Transformation:** The sources were converted into the frequency domain using a DFT algorithm.

3.a   **Acoustic Filtering (Optional):** The acoustic contribution to the total pressure fluctuation was filtered using the Acoustic Perturbation Equation (APE).

4.a   **SPL Propagation:** The SPL, separated from the PFL using APEs, was calculated over the vehicle's surface and in the surrounding space (dependent on Step 3.a).

3.b   **PFL Calculation:** The PFL was calculated directly on the vehicle's surface by propagating unfiltered sources (an alternative to Steps 3.a and 4.a).

It is important to note that Step 3.b served as an alternative to Steps 3.a and 4.a, particularly when filtering of the acoustic contribution was deemed unnecessary for the analysis. A schematic representation of the computational steps used to model the exterior aeroacoustic field is provided in **Figure 4.9**.

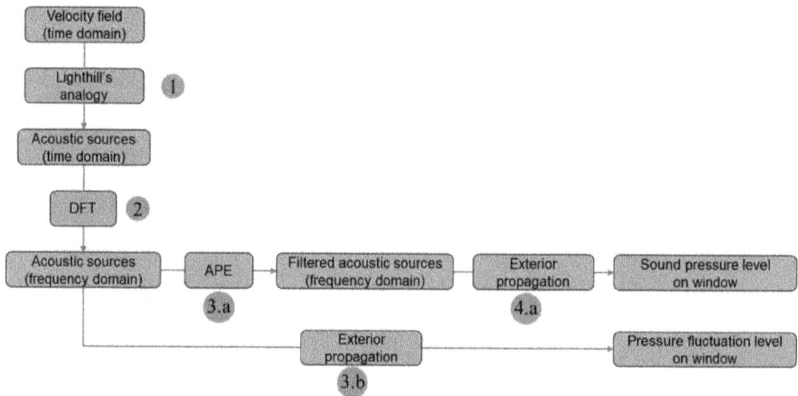

Figure 4.9:   Numbered blocks illustrating the computational steps involved in modeling the exterior acoustic field

4.4   Noise transmission into the vehicle cabin

The simulation of the noise transmission into the cabin was based on the following key assumptions:

- The windows (side and rear) are the only flexible components of the vehicle, while the rest of the body is considered infinitely rigid.
- The cabin is assumed to be completely airtight, with the influence of the sealing system deemed negligible.
- The acoustic contribution to the total pressure fluctuation is considered the dominant factor in noise transmission into the cabin [37].

The acoustic domain for interior noise propagation is defined by the geometry of the windows and the interior cabin. For computational efficiency, only one-quarter of the cabin (driver's region) was analyzed for the side mirror region, and half of the cabin (rear passenger's region) was analyzed for the roof spoiler region. These areas, highlighted in dark gray in **Figure 4.10**, represent a simplified modeling approach aimed at optimizing the simulation for localized component analysis while concurrently minimizing computational costs. The experimental setup described in Chapter 5 was designed to follow these simplifications.

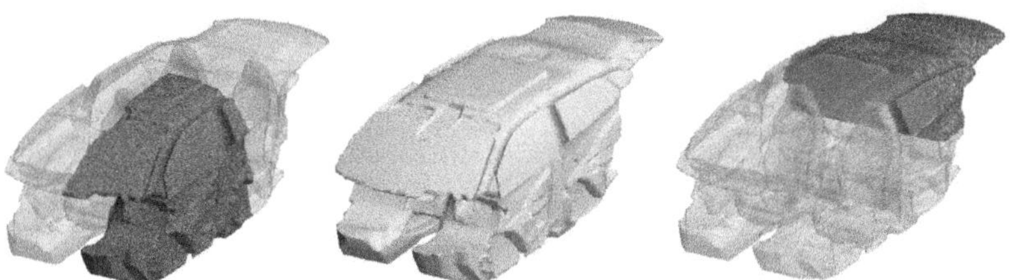

Figure 4.10: One-quarter cabin (driver's region, left), and half-cabin (rear passenger's region, right) used to define the acoustic domains for the interior noise modeling

In the third and last step of the CAA workflow, the SPL at the passenger's ear location was calculated. This was achieved by implementing the SPL distribution on the side and rear windows, computed in the second step (see chapter 2.1.3), and the eigenmodes previously determined for these windows, as the boundary conditions of the model (see **Figure 4.11**). The material properties of the glass (young modulus (E), thickness (h_s), density (ρ_s) and the Poisson ratio (ε_s)) are listed in **Table 4.2** for the vibro-acoustic simulation.

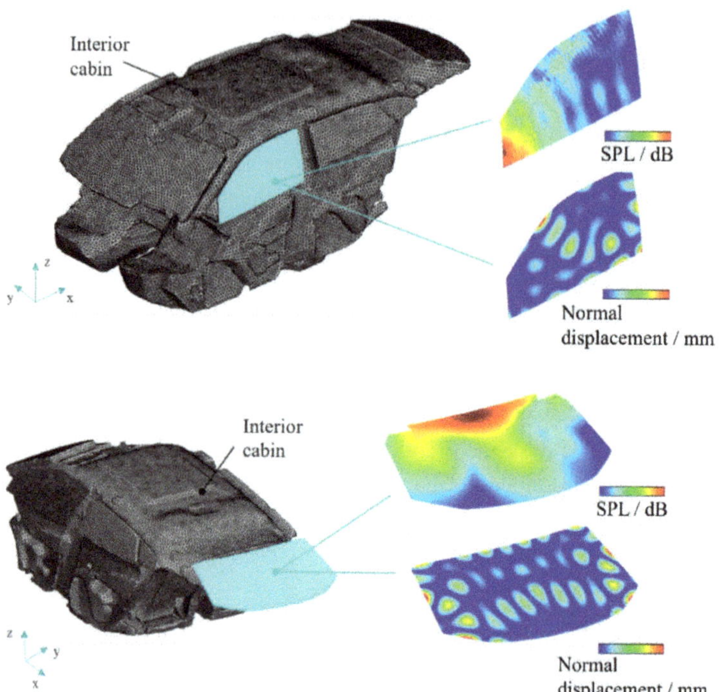

Figure 4.11:   Example of boundary conditions for the Vibro-Acoustic model: side mirror region including the side window (top) and roof spoiler region including the rear window (bottom)

Table 4.2:   Glass properties of the side and rear windows

Glass properties	Side Window	Rear Window
Young modulus (E) / Pa	6e+10	6e+10
Poisson ration (ε_s) / -	0.23	0.23
Solid density (ρ_s) / Kg/m^3	2500	2500
Glass thickness (h_s) / mm	4.85	3.50

A modal analysis was performed on the window as an unconstrained structure, omitting the effects of the clamping system [52]. To account for the absorption properties of the vehicle's interior cavity, the reverberation time (RT_{60}) inside the cabin was measured. This was achieved by recording the sound decay over 60 dB after switching off a broadband noise source emitted by a loudspeaker positioned in the respective cavity (see **Figure 4.12** left). Using this approach, the frequency-dependent sound absorption coefficient ($\phi_c(f)$) (**Figure 4.12** right), was determined based on Eq. 4.6, as described in [83], This coefficient was then incorporated into the finite element (FE) model to simulate the acoustic behavior of the inner cavity with greater accuracy.

$$\phi_c(f) = \frac{\ln(10^6)}{RT_{60} \cdot 2\pi f} \qquad\qquad \text{Eq. 4.6}$$

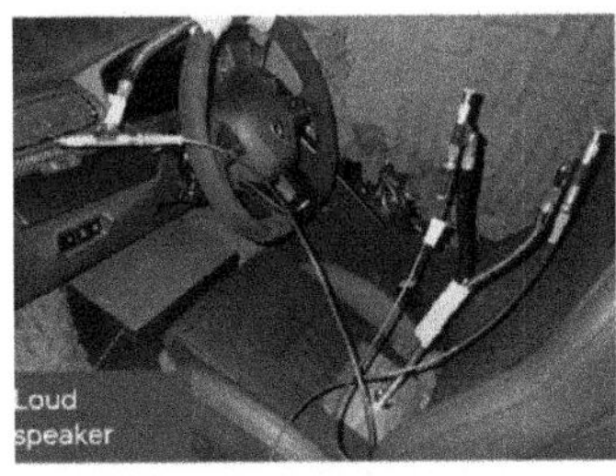

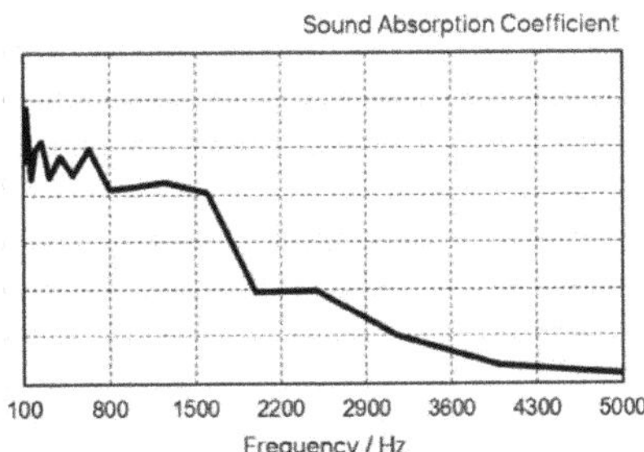

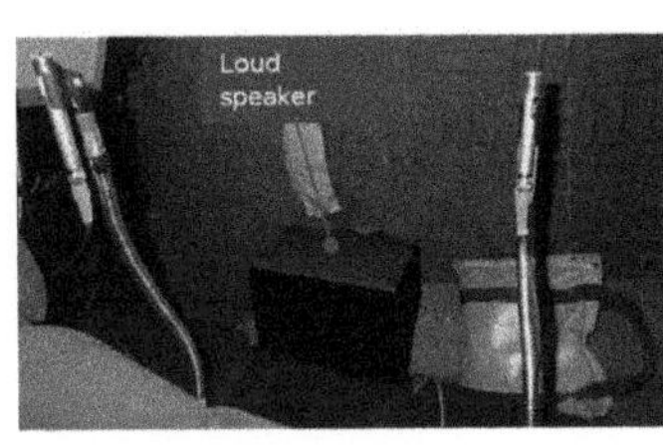

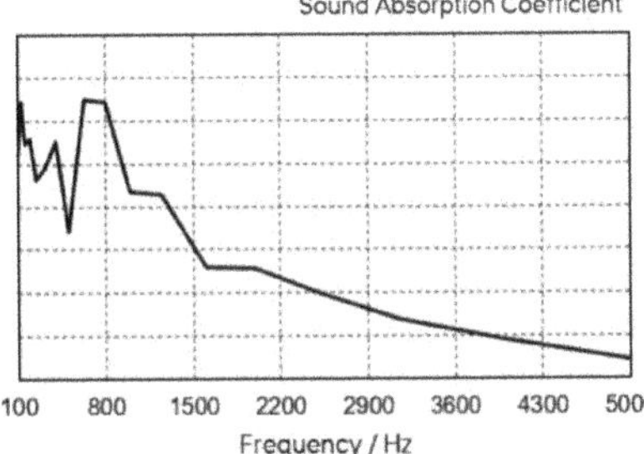

Figure 4.12: Loudspeaker setup and measured sound absorption coefficient for the driver's region (top) and the rear passenger's region (bottom)

5 Validation of the Numerical Model

In this chapter, a comprehensive overview of the step-by-step validation process is presented. The experimental setup, meticulously designed to investigate both exterior and interior noise phenomena, is then described in detail. Finally, an in-depth discussion focuses on the validation of the numerical process, highlighting its application across distinct regions of the vehicle.

5.1 Validation process overview

The numerical tool was validated by comparing its predictions against wind tunnel experimental data. The analysis focused on the side mirror and roof spoiler regions to evaluate the accuracy and effectiveness of the developed numerical approach for aeroacoustic analysis. This validation process was conducted in accordance with the sequential steps of the computational aeroacoustics workflow illustrated in **Figure 4.2**. An outline of the validation methodology for each stage is illustrated in **Figure 5.1**.

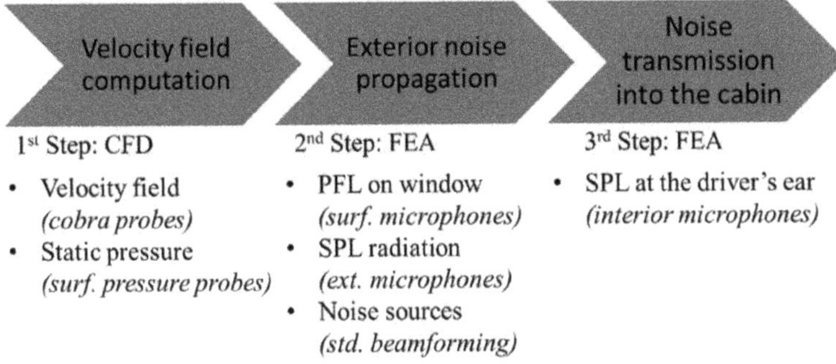

Figure 5.1:   Step-by-step validation process

First, CFD predictions were validated against experimental flow field data, focusing on the velocity field within the turbulent wake of the studied components and the static pressure distribution on the vehicle surface. Subsequently, the FE simulations were compared with aeroacoustic measurements, including the PFL on the windows, the SPL in the far field, and acoustic sources identified through beamforming techniques. Lastly, the VA model was evaluated by comparing the predicted SPL at the passengers' ear with interior cabin measurements.

This structured validation approach rigorously benchmarked the numerical predictions against experimental results at every stage, demonstrating their accuracy and reliability.

5.2   Experimental setup

The experimental investigation was conducted entirely in the full-scale aeroacoustic wind tunnel of the University of Stuttgart, operated by FKFS. To ensure consistency and repeatability, measurements were performed on two vehicles of the same model and configuration. All data were collected at a speed of 140 km/h and a yaw angle of $0°$, aligning with the simulation conditions. As is typical in aeroacoustic studies, wheel rotation and ground simulation were excluded, with the rolling road and boundary layer suction systems not employed. A dedicated setup was designed for both the vehicle exterior and interior to ensure comparability between experimental and numerical results.

5.2.1   Setup and measurement devices for vehicle exterior

The numerical tool is designed to predict both exterior and interior noise originating from components and specific geometric features of the vehicle. In this investigation, the focus was on the side mirror and roof spoiler, which were modeled in simulations to analyze how noise generated in these regions transmits through the side and rear windows, respectively, into the cabin. By concentrating solely on these exterior design features, potential noise sources such as those from sealing systems and gaps between components were intentionally excluded to ensure clarity in evaluating the impact of the geometry itself. To align with these simulation assumptions, experiments were performed using a fully taped vehicle (see **Figure 5.2**) with a closed cooling system, following standard practices in exterior aeroacoustic vehicle development.

Figure 5.2:   Fully taped configuration of the Lamborghini Urus with a closed cooling system in the wind tunnel test section (left), and a close-up view of the fully taped side mirror region (right)

The flow field was measured using cobra probes from Turbulent Flow Instrumentation (TFI), positioned in the wake of the side mirror and roof spoiler. Static pressure measurements were obtained using surface pressure probes mounted along the centerline of the vehicle on the hood and roof. (**Figure 5.3**).

Figure 5.3:   Cobra probes mounted on a traversing gear positioned in the side mirror wake (left), and surface pressure probes installed on the vehicle hood (right)

The PFL was measured using surface microphones flush-mounted on the side and rear windows (**Figure 5.4** top). The SPL in the far field was captured by exterior microphones positioned near the vehicle but outside the turbulent region to exclude significant hydrodynamic contributions. These microphones, equipped with nose cones, were mounted on a streamlined traversing gear to minimize their impact on the flow field and the recorded noise (**Figure 5.4** center). Acoustic sources were identified using standard beamforming techniques with a microphone array system [93] (**Figure 5.4** bottom).

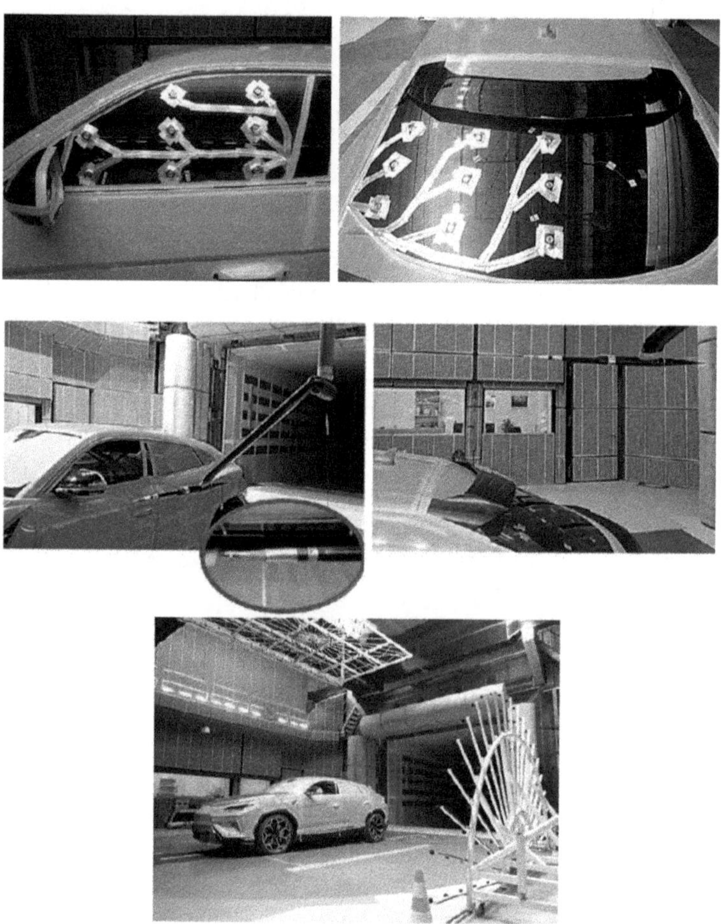

Figure 5.4: Flush-mounted surface microphones (top), exterior microphones with nose cone (center) and microphone array system in the aeroacoustic wind tunnel (bottom)

5.2.2 Setup and measurement devices for vehicle interior

To focus on noise transmission through the side and rear windows, the interior cabin setup was designed to attenuate noise from other areas, such as the windshield, underfloor, and additional windows. The interior was divided into isolated spaces, as shown in **Figure 5.5**, enabling separate investigations of the side mirror and roof spoiler effects on the driver and rear passengers, respectively. A double-layer of sound absorption foam was used to separate the driver's area from the rear passenger seats, ensuring independent analysis of noise generated in each region and eliminating cross-region noise interference.

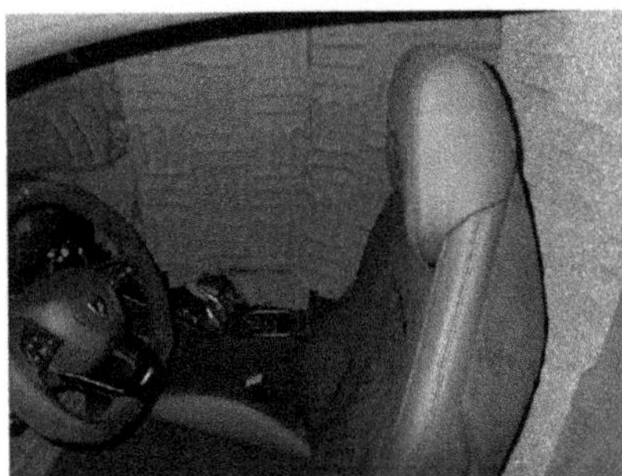
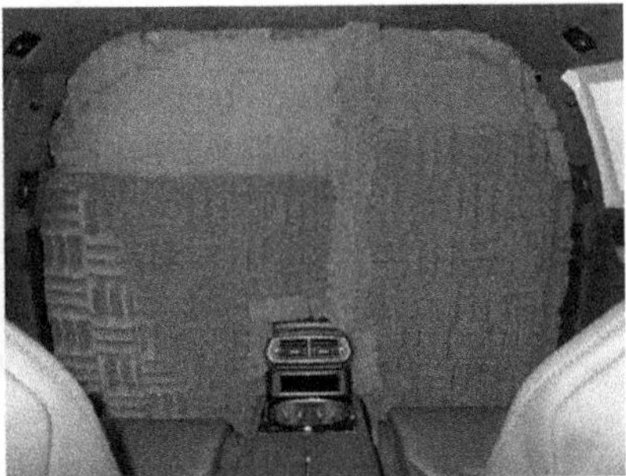

Figure 5.5: Insulation strategy separating the driver's seat (left) and the rear passenger seats (right) from the rest of the cabin using sound-absorbing materials

The windshield and non-investigated windows were internally insulated with a two-layer system: a heavy material layer to dampen glass vibrations and an absorbing material layer to reduce interior noise radiation (**Figure 5.6** left). A similar strategy was applied to the underbody, typically causing dominant low-frequency noise, using sandbags and absorbing foam inside the cabin (**Figure 5.6** right). Additionally, foam insulation was added to the trunk area to eliminate unwanted noise contributions, as the trunk was excluded from the VA model to reduce computational costs.

Figure 5.6:   Two-layer insulation installed on the windshield (left) and on the cabin floor (right) for noise attenuation

The noise transmitted into the cabin was measured using microphones placed on the driver and rear passenger seats, including standard and directional types. Additionally, reference microphones connected to the external array system were installed in both cabin compartments for correlation with exterior wind noise sources (see **Figure 5.7**). All the acoustic measurements were conducted over a time window of 30 seconds to ensure robust sampling and mitigate unsteady flow effects. The same DFT algorithm used in the numerical model was applied for frequency-domain post-processing of the experimental data.

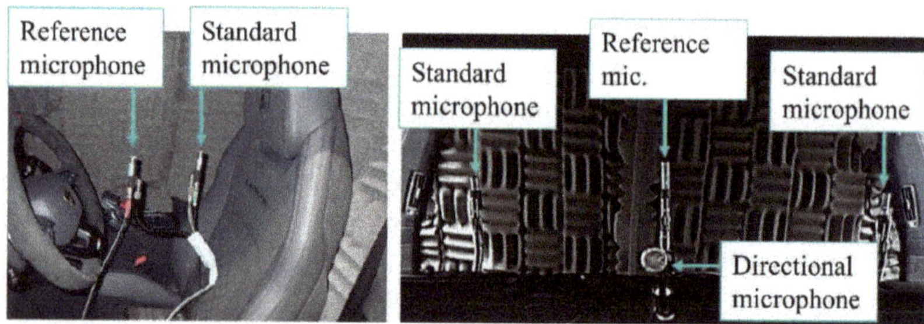

Figure 5.7:   Standard and reference microphones positioned on the driver's seat (left) and on the rear passengers' seats (right) for interior noise measurements

5.3   Validation of the side mirror and roof spoiler regions

Given the sequential nature of the simulation process - where each computational step provides input for the next - it is critical to assess how assumptions, simplifications, and inaccuracies at one stage may propagate and affect subsequent modeling steps. For this reason, a step-by-step validation is essential, not only to verify the numerical model's accuracy but also to identify potential challenges and limitations at each computational stage.

5.3.1   Suitability of the CFD simulations for aeroacoustic analysis

The validation of the CFD predictions is based on time-averaged analysis. Simulation results were averaged over 0.5 seconds (refer to **Figure 4.5**), while the experimental data were averaged over 30 seconds. The velocity field was measured in the wake of the side mirror and in the wake of the roof spoiler using grids with 42 and 54 measurement points, respectively, defined on Y-Z planes. **Figure 5.8** and **Figure 5.9** show a direct comparison between the CFD predictions (light grey line) and the experimental results (black diamonds). The plots display the normalized velocity magnitude (U_{mag}), relative to the free stream velocity. Each data point corresponds to a location along the selected row, marked by orange dots on the left and highlighted by a white arrow. For brevity, only the grid rows capturing the core of the component's wake are presented as representative examples.

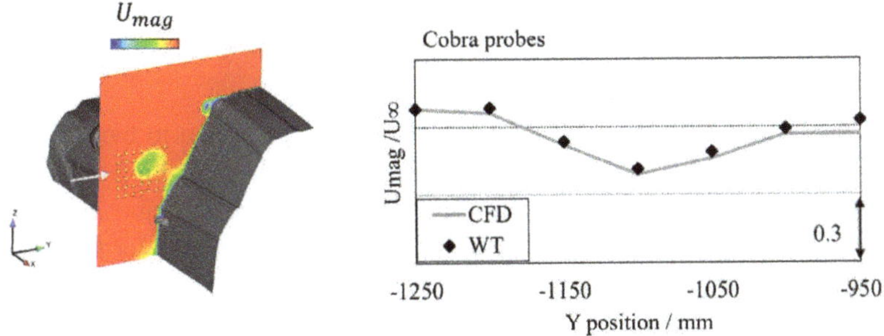

Figure 5.8:   Velocity field validation for side mirror region: CFD predictions vs WT measurements

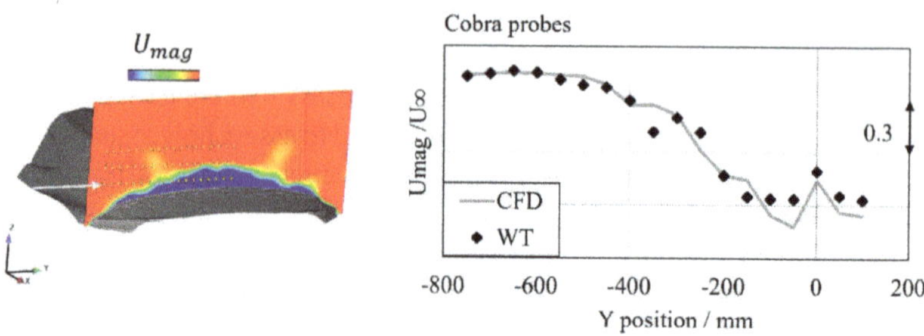

Figure 5.9:   Velocity field validation for roof spoiler region: CFD predictions vs WT measurements

A strong agreement was observed between CFD predictions and experimental results, with percentage errors ($\varepsilon_\%$) (as defined in Eq. 5.1) below 6%, except for some higher local deviation in the roof spoiler highly turbulent regions. These discrepancies can be attributed to factors such as CFD grid resolution, turbulence modeling limitations, and the high-Reynolds approach, which may introduce inaccuracies in areas with significant flow separation. Additionally, cobra probes used for measurements, with their specific measurement cone (see Section 3.2.2), have limitations in capturing velocity components in high turbulent or back-flow regions, potentially contributing to the observed deviations.

$$\varepsilon_\% = \frac{|Simulation\ value - Experimental\ value|}{Experimental\ value} \cdot 100 \qquad \text{Eq. 5.1}$$

Despite these challenges, the transition from the free stream to the turbulent wake was accurately predicted, and regions not dominated by high separation showed high accuracy. Overall, the CFD models provided reliable and accurate velocity field data suitable for use as input in the FE model for the subsequent step of the CAA process.

An additional validation was conducted to assess the accuracy of the wall model employed in the DDES. This was achieved by comparing the simulated static pressure coefficient ($c_{P,stat}$) with experimental measurements taken along the vehicle centerline at its surface, specifically on the hood and roof.

Figure 5.10 presents a direct comparison between the CFD simulation (light grey line) and wind tunnel measurements (black diamonds). The static pressure coefficient is plotted as a function of the X-coordinate, with the origin aligned with the front wheel axle.

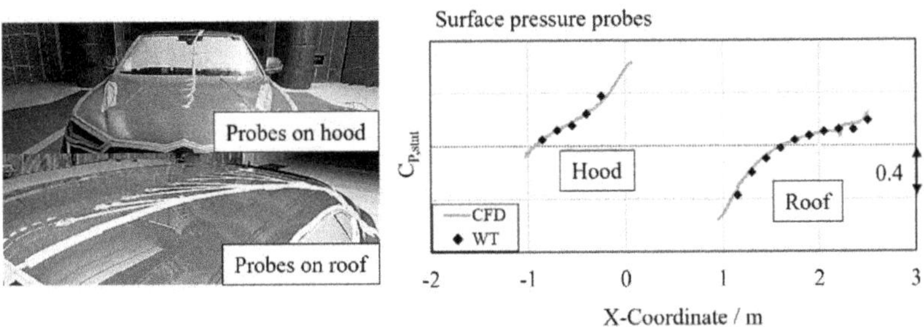

Figure 5.10:　　Validation of the static pressure coefficient along the vehicle centerline: CFD predictions vs WT measurements

The modeled static pressure coefficient closely matched the measured data, with an average deviation of approximately 10 counts on the hood and 20 counts on the roof, where 1 count corresponds to 0.001 of $c_{P,stat}$. These results validated the suitability of the implemented wall model and boundary layer discretization for capturing the flow field, reinforcing the overall reliability of the numerical model.

5.3.2　　Analysis of the aeroacoustic sources and exterior propagation of noise

To ensure consistency and repeatability, measurements were performed on two vehicles of the same model, referred to as Urus 1 and Urus 2. **Figure 5.11** compares the PFL on the side window and the SPL in the far field, both measured at exactly the same locations relative to the vehicle using the same measurement procedure. The PFL was recorded by surface microphones on the side window, while the SPL was captured by exterior microphones positioned at different locations 3 meters far from the car centerline. For brevity, results from one surface microphone and one exterior microphone are shown.

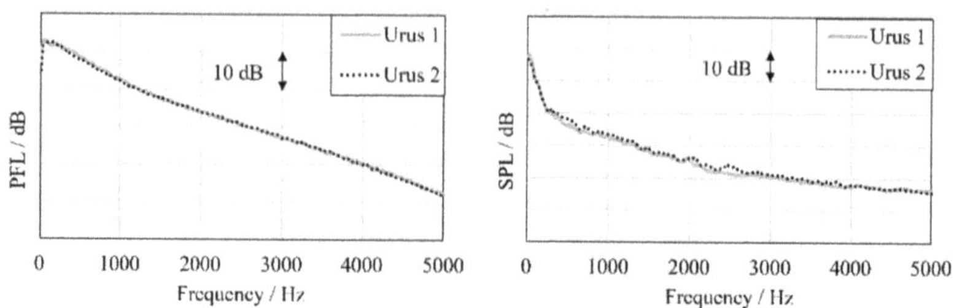

Figure 5.11:   Comparison of measurements on two identical vehicles: PFL
on the side window (left) and SPL at far-field (right)

The comparisons demonstrated high measurement consistency, with both vehicles exhibiting similar turbulent flow conditions and noise sources across the
vehicle.

The predicted pressure fluctuation levels (PFL) on the side and rear windows
were validated using data from 8 surface microphones mounted on the side
glass and 9 on the rear glass (as shown in **Figure 5.4** top). **Figure 5.12** and
Figure 5.13 present a direct comparison between simulation results (FEM)
and experimental data (WT). The comparison focuses on three representative
microphones on the side window (positions 3, 4, and 5) and three on the rear
window (positions 9, 10, and 11).

The PFL predicted by the FE simulation demonstrated good agreement with
experimental data, though certain deviations are observed in both the side mirror and roof spoiler regions. A broadband overestimation of PFL was noted
for microphones 4 and 5 near the side mirror, as well as microphones 9 and 10
near the roof spoiler, up to 2 kHz. These discrepancies can be linked to common factors affecting the numerical accuracy across both regions.

The flat probe geometry was omitted in the computational CFD and FE models, neglecting two critical effects. First, the hydrodynamic contribution to
pressure fluctuations, with its shorter wavelengths relative to acoustic fluctuations, becomes averaged over the finite surface of the microphone membrane.
This averaging reduces experimental sensitivity to high-frequency hydrodynamic fluctuations but is absent in the numerical model, potentially leading to
an overestimation of high-frequency PFL.

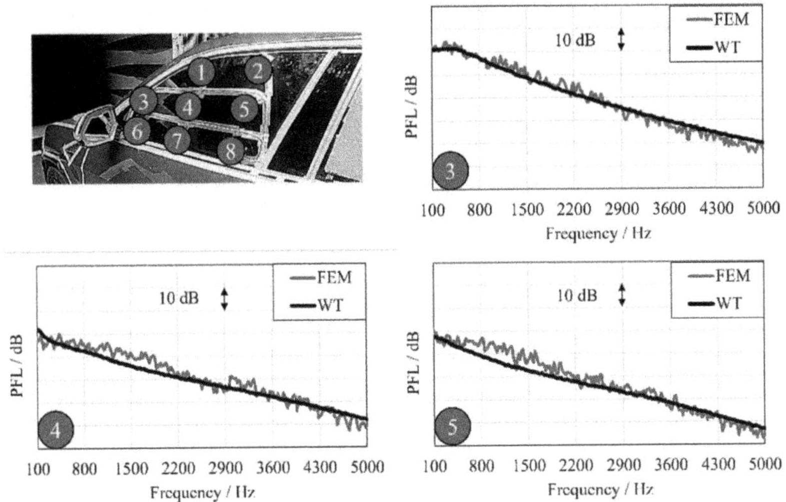

Figure 5.12: Validation of the PFL induced on the side window: FEM predictions vs WT measurements

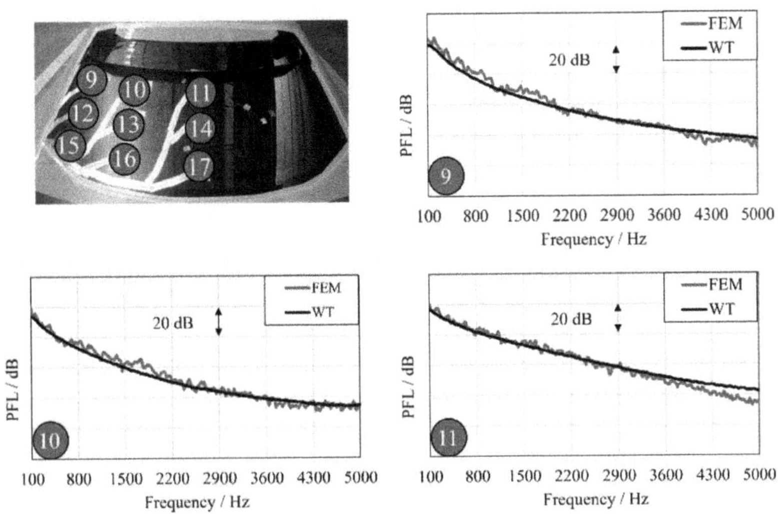

Figure 5.13: Validation of the PFL induced on the rear window: FEM predictions vs WT measurements

Conversely, neglecting the flow distortion caused by microphone placement in the experiments introduces additional pressure fluctuations, contributing to further inconsistencies. A slight underestimation of high-frequency PFL (above 4 kHz) was also observed for both regions, particularly for microphone 11 near the rear window. This underestimation can be attributed to limitations of the numerical model, including the 2 mm CFD mesh size, which might be limited in resolving small hydrodynamic wavelengths, and the wall functions employed for boundary layer modeling. These wall functions are less effective in areas of significant flow separation, such as the wake regions downstream of the roof spoiler and near the side mirror.

Nevertheless, the numerical models accurately capture the experimental trends. Incorporating exact probes geometry could address flow distortion and hydrodynamic averaging, the likely origin of observed deviations. However, given the satisfactory accuracy of the acoustic results, as will be detailed later, further refinements to the total pressure fluctuation modeling were deemed unnecessary.

Further investigation focused on sound propagation from the side mirror and roof spoiler regions into the far field. Acoustic pressure fluctuations were analyzed using microphones placed near the wakes of these regions. To ensure reliability and avoid any disturbance, wind tunnel background noise was assessed in the empty test section at the same microphone positions: 18, 19, 20 for the side mirror (**Figure 5.14**, top-left) and 21, 22, 23 for the roof spoiler (**Figure 5.15**, top-left). For positions 18, 19, 20, background noise was significantly lower than the measured SPL across all frequencies, confirming reliable validation for the side mirror region up to 5 kHz. However, for positions 21, 22, 23, background noise above 2 kHz was comparable to the SPL generated by the roof spoiler, limiting validation in this region to frequencies below 2 kHz.

At this validation step, the positions of the exterior microphones were chosen in order to measure the acoustic contribution only. Therefore, the microphones were placed outside the turbulent region, in order to minimize hydrodynamic fluctuation in the data, although close enough to the side mirror and the roof spoiler in order to be able to neglect additional noise sources coming from other regions of the vehicle. The SPL predicted by virtual microphones and measured in the experiments is shown in the following diagrams (**Figure 5.14** and **Figure 5.15**).

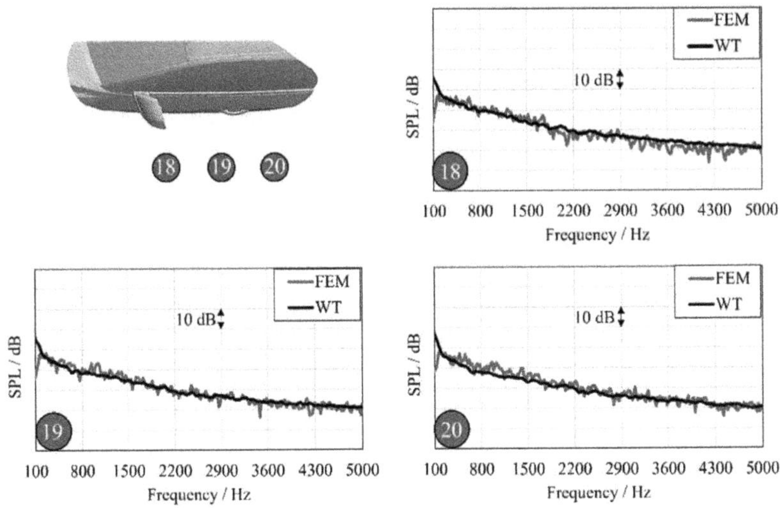

Figure 5.14: Validation of the SPL exterior propagation from the side mirror region: FEM predictions vs WT measurements

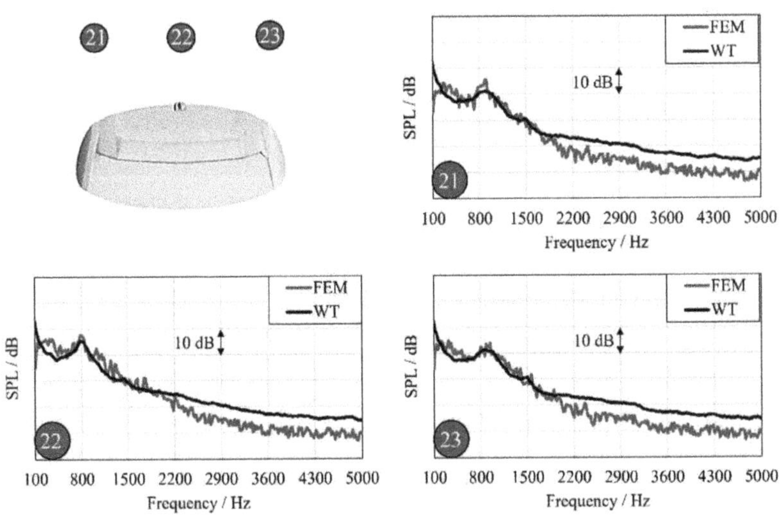

Figure 5.15: Validation of the SPL exterior propagation from the roof spoiler region: FEM predictions vs WT measurements

The simulation predictions closely align with experimental results. A notable difference in amplitude fluctuations between the numerical model and wind tunnel measurements arises from the differing time lengths used in the simulations (0.2 s) and experiments (30 s). However, this discrepancy does not compromise the validation's purpose or the numerical tool's effectiveness in vehicle development.

The low-frequency deviations observed below 200 Hz in both regions, as explained in Chapter **Error! Reference source not found.**, are acceptable, as the significant aeroacoustic effects primarily occur at higher frequencies. Additionally, deviations above 2 kHz were expected for the roof spoiler microphones, where the CAA underpredicts noise due to SPL levels falling below the background noise at these locations. It is important to note that the CFD model relied on a simplified simulation domain (see **Figure 4.3**) that excludes wind tunnel geometry and features, thereby omitting any background noise effects.

Overall, the developed FE model was deemed suitable for accurately modeling the exterior acoustic field and providing reliable input for VA simulations, particularly for PFL and SPL distributions on vehicle windows.

5.3.3   Noise transmission into the vehicle cabin

The validation process was concluded with an analysis of the VA model's prediction of noise transmission into the vehicle cabin. The robustness and repeatability of the experimental results were confirmed by measurements from both Urus 1 and Urus 2. Interior noise was recorded using microphones placed at exactly the same positions inside the cabins at the driver's ear. A comparison of these measurements is presented in **Figure 5.16**, demonstrating consistency across both vehicles, as similarly illustrated in **Figure 5.11**.

In the final step of the simulation process, the PFL distribution on the side and rear windows, obtained from the FE model, served as a boundary condition for the VA model. This input was combined with the eigenmodes of the windows to accurately predict interior noise propagation. The validation was conducted by comparing the SPL at the driver's and rear passenger's ear, measured using interior standard microphones (refer to **Figure 5.7**), with the simulated SPL at virtual microphones positioned at the same locations. **Figure 5.17** presents the interior SPL validation for both the driver and rear passenger, induced within the side mirror and the roof spoiler regions, respectively.

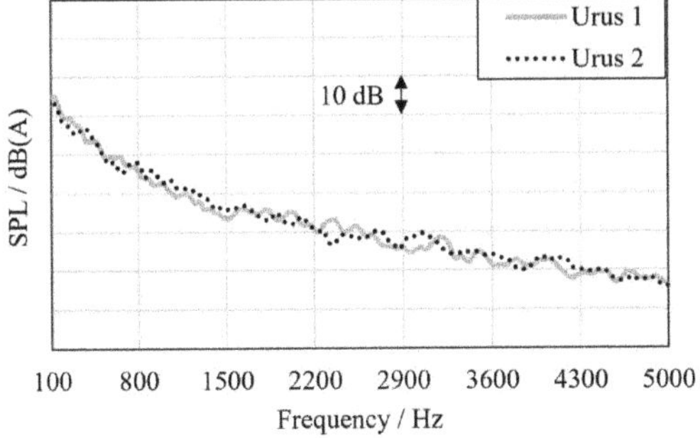

Figure 5.16:　Comparison of interior noise measurements on two identical vehicles: SPL at the driver's ear

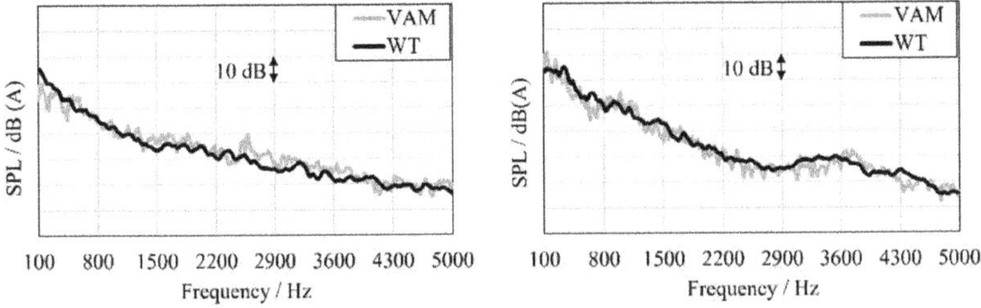

Figure 5.17:　Validation of SPL at the driver's ear (induced from side mirror region, left) and rear passenger's ear (induced from roof spoiler region, right): VAM predictions vs WT measurements

For the side mirror region, a strong correlation was observed between simulation predictions and experimental results. While slight deviations were observed at low frequencies (100 - 280 Hz) and in the proximity of the side window's coincidence frequency (c.a. 2500 Hz), accuracy was high in the rest of the range. For the roof spoiler region, the numerical simulation also closely matched the interior noise measurements. Despite overall accuracy, in both cases slight deviations occurred at frequencies between 2 kHz and 4 kHz. The high-frequency deviations may result from reflection phenomena within the cabin. These effects, coupled with limitations in modeling the shading effects caused by trim work and relying solely on the sound absorption coefficient, could contribute to the observed discrepancies. The potential causes of the observed deviations will be further examined in Chapter **Error! Reference source not found.**.

Overall, the simulations demonstrated exceptional accuracy across all three steps of the process, confirming their applicability in vehicle development. The model proved highly effective for analyzing unsteady flow fields, modeling exterior noise, identifying noise sources, and predicting interior noise induced by specific regions and components, such as the side mirror and roof spoiler. This comprehensive validation underscores the robustness and reliability of the developed numerical model. Furthermore, it establishes the groundwork for the next chapter, where implications, limitations, and potential refinements of the approach will be addressed.

Notably, this work presents a step-by-step validation of the hybrid numerical model - from CFD to FEM and VAM - ensuring consistency across simulation domains. The validation of the roof spoiler region represents a key achievement, as no prior studies have publicly addressed this area, likely due to confidentiality constraints. The effective experimental setup and accurate simulation of this complex region not only demonstrate the model's robustness but also offer practical guidance for simulating and validating similar scenarios. Moreover, the validation framework, including the experimental methodology, is generalizable and can be applied to other hybrid numerical models based on the Navier-Stokes equations, regardless of the specific CFD or FEM solvers used. Together, these contributions address a significant gap in the state of the art and advance the methodology for aeroacoustic model validation.

6 Discussion of Results

This chapter presents a detailed discussion of the results, emphasizing the simulation outcomes to identify limitations and areas for improvement related to both the numerical model and the experimental setup. In addition, a combined analysis of CAA simulations and wind tunnel measurements is conducted to investigate the mechanisms of wind noise generation in the side mirror and roof spoiler regions. Notably, the source analysis of the roof spoiler addresses a gap in the existing literature, as no scientific studies have been published on this topic. The integrated approach presented here demonstrates the value of combining numerical and experimental methods to achieve a more comprehensive understanding of aeroacoustic phenomena in complex vehicle components.

6.1 Considerations on the aeroacoustic simulation process

The numerical process relied on simplifying assumptions at each step, carefully designed to balance accuracy and computational efficiency, making the simulations suitable for vehicle development applications. These key assumptions were hence analyzed to evaluate their impact on the overall simulation outcomes and to identify potential criticalities arising from these simplifications.

6.1.1 Further analysis of the numerical model at low frequencies

Low-frequency noise inside the cabin, typically caused by underbody pressure fluctuations and tire rolling noise, is often overlooked in studies focusing on vehicle's geometry features and aerodynamic appendices such as side mirrors and roof spoilers. Numerical research in the literature tends to prioritize higher frequency ranges, deemed more relevant to passenger noise perception, while discrepancies below 500 Hz are frequently reported but rarely analyzed in detail. This aspect represents a notable gap in the current state of the art. Therefore, the present analysis includes a focused examination of the low-frequency content to evaluate the limitations of current modeling approaches and to ensure that potentially relevant contributions are not disregarded.

A detailed analysis was conducted to better understand the simulation process at low frequencies (100 – 500 Hz range) and to investigate the deviations observed in the side mirror case. This analysis examined the sensitivity of the simulation to physical time and the boundary conditions applied to the window, including the distribution of total and acoustic pressure fluctuation on the glass. The side mirror region served as the reference for all these investigations.

The numerical model requires a sufficient number of recorded wavelength cycles to accurately capture acoustic pressure fluctuations. Consequently, the minimum physical time required is determined by the largest wavelength corresponding to the lowest target frequency - in this study, 100 Hz. This establishes the minimum time necessary to model the desired frequency range. Therefore, to assess the impact of the physical time used for extracting the unsteady velocity field from CFD (see **Figure 4.5**), the same simulation was conducted with varying time lengths (0.1 s, 0.2 s, 0.3 s, and 0.4 s). **Figure 6.1** presents a comparison of the predicted interior noise at the driver's ear for these time spans against experimental data, highlighting the importance of selecting an appropriate time window for accurate CFD velocity extraction.

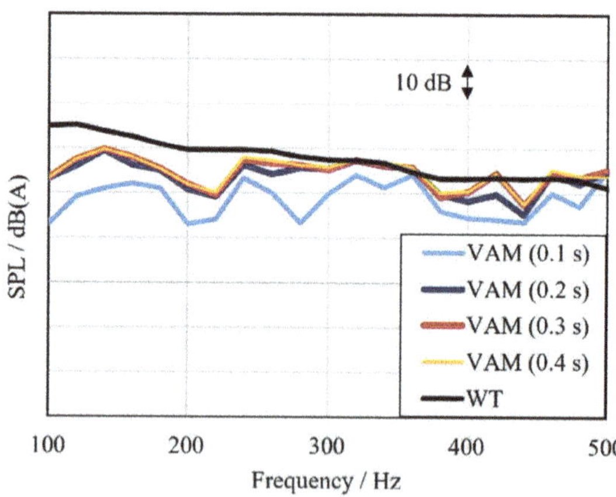

Figure 6.1:   Sensitivity analysis of velocity extraction time: comparison of SPL at the driver's ear from vibro-acoustic model predictions (VAM) and wind tunnel measurements (WT)

The graph illustrates the impact of increasing physical time on low-frequency simulation predictions. At 0.1 s, significant SPL underestimation occurs. Convergence is achieved at 0.2 s, with only minor improvements observed at 0.3 s and 0.4 s, which would not justify the higher computational cost. Thus, 0.2 s is confirmed as an optimal balance between accuracy and efficiency for low-frequency aeroacoustic analysis.

Furthermore, the influence of the pressure distribution applied as a boundary condition on the side window was investigated. As outlined in Section 4.4, the VA model assumes a free-vibrating side window, neglecting the sealing and clamping systems. To evaluate this simplification, interior noise predictions were also analyzed with zero displacement imposed at the window edges. **Figure 6.2** compares the predicted interior noise (SPL at the driver's ear) for both cases – free-vibrating and clamped window - under the same wind load, against experimental data.

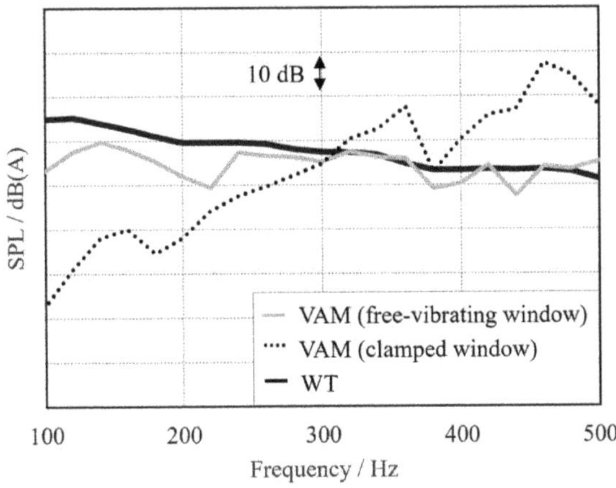

Figure 6.2: SPL at the driver's ear: VAM prediction with free-vibrating window vs VAM prediction with clamped window vs WT measurements

When adopting a free-vibrating window model, the predicted interior SPL (light grey line) shows minor deviations below 250 Hz. However, clamping the window edges leads to greater underestimation at frequencies below 300 Hz, while significant overestimation occurs at higher frequencies (black dotted line). To investigate the causes of this behavior in the VA model, the low-frequency eigenmodes of side window under different boundary conditions (free-vibrating and clamped) were analyzed, as illustrated in **Figure 6.3**.

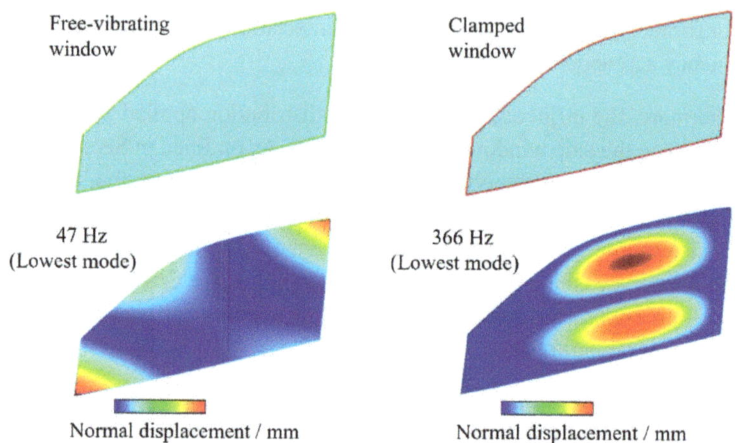

Figure 6.3:   Side window and lowest eigenmodes calculated under different boundary conditions: free-vibrating (left) and clamped (right)

A free-vibrating window exhibits six rigid modes and additional 12 modes up to 500 Hz, with the lowest mode occurring at 47 Hz (**Figure 6.3** bottom-left). In contrast, a clamped window has only 2 modes within this range, with the lowest at 366 Hz (**Figure 6.3** bottom-right). The absence of low-frequency modes in the clamped configuration leads to underestimation of interior noise below 300 Hz, as no modes are excited, reducing the interaction between the PFL distribution on the window and the modal response. Moreover, the clamped window configuration leads to larger vibration amplitudes and a concentrated normal displacement at the window's center, as the clamped edges enforce zero displacement at the borders. This BC results in a significant overestimation of interior noise between 300Hz and 500Hz (as shown in **Figure 6.2**) due to an exaggerated response of the window vibration on the external wind load. Consequently, a free-vibrating window is the most appropriate assumption when sealing and clamping systems are excluded from the model, as proven by the accurate results presented in **Figure 6.2**. On the contrary, fully clamped window should be avoided in this scenario.

Lastly, the impact of both acoustic and hydrodynamic contributions to the total pressure fluctuation distribution on the window was analyzed. The simulation process was repeated using the SPL distribution on the side window, filtered by the APE (see Section 4.3), in place of the PFL distribution. Results for both boundary conditions (SPL and PFL on the window) and the corresponding wind tunnel experiment are presented in **Figure 6.4**. Given the broader implications of these findings, discussed further in the following sections, the results are shown across the entire frequency spectrum (100 - 5000 Hz) for both scenarios - the side mirror and the roof spoiler. A zoomed-in view of the low-frequency range is also provided.

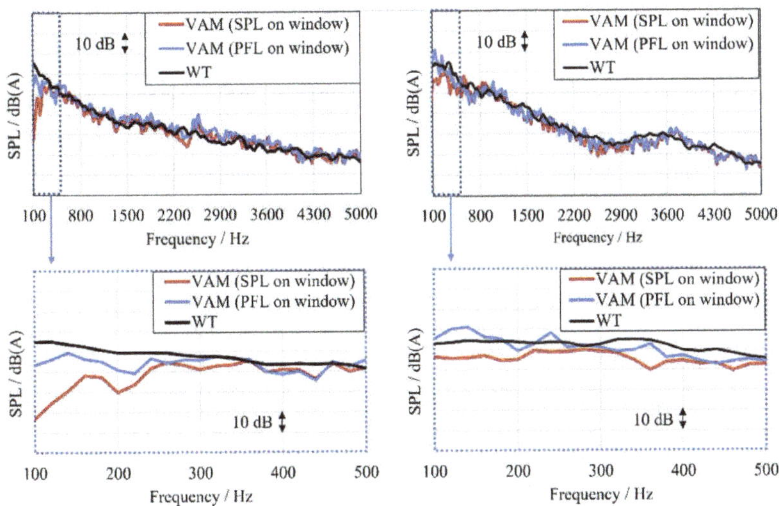

Figure 6.4: SPL at the driver's ear (left) and rear passenger's ear (right) under different wind load conditions (SPL and PFL): VAM predictions vs WT measurements

The predicted interior SPL closely aligns with experimental results across the entire frequency range, regardless of the boundary condition applied (i.e., SPL or PFL distribution on the window). Above 360 Hz, filtering the acoustic contribution does not significantly affect the SPL prediction at the passengers' ear. However, below 360 Hz, the model's accuracy decreases notably, suggesting that hydrodynamic contributions may dominate noise transmission at these frequencies. Thus, relying solely on the SPL in the VA model may not suffice for accurate low-frequency predictions.

Previous studies [37, 88, 89, 95] have demonstrated that the acoustic contribution to total pressure fluctuations plays a dominant role in cabin noise transmission for standard road vehicles. This dominance is attributed to the similarity between the scales of acoustic fluctuations and the structural bending wavelengths of the window, leading to efficient transmission through interaction with the window's eigenmodes. Conversely, hydrodynamic fluctuations, despite their higher pressure amplitudes, have smaller wavelengths that are typically damped by the glass, making them less likely to transmit into the cabin.

The simulation results presented here confirm this broad-band behavior. For a better understanding of the phenomenon discussed, the wavenumbers related to the acoustic fluctuation (k_a), to the hydrodynamic fluctuation (k_h) and to the structural bending wavelength of the window (k_b) have been calculated for both windows (side and rear), following the procedure presented in [88], see **Figure 6.5**.

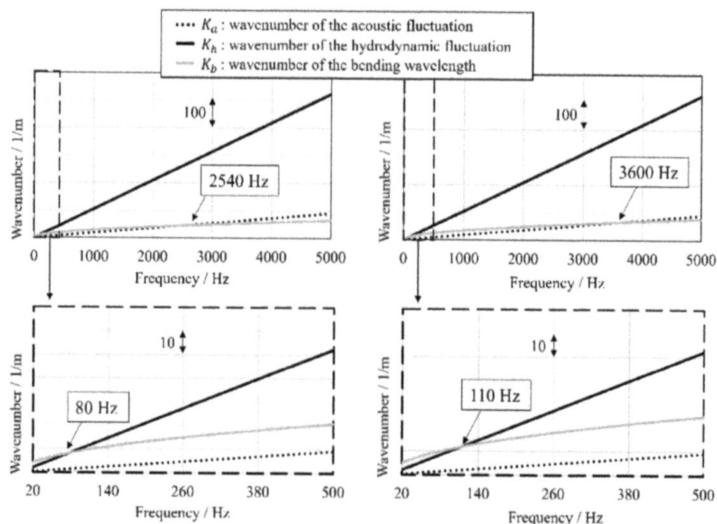

Figure 6.5:   Wavenumbers related to acoustic fluctuation, hydrodynamic fluctuation, and structural bending wavelengths of the side window (left) and rear window (right)

The acoustic wavenumber (black dotted lines) aligns closely with the glass bending wavelength (grey line) across most of the frequency range (360 Hz - 5000 Hz) for both the side and rear windows (see **Figure 6.5**, top charts). Notably, acoustic coincidence frequencies - where the acoustic and glass wavenumbers intersect - occur at approximately 2540 Hz for the side window and 3600 Hz for the rear window. This alignment explains the interior noise behavior observed in **Figure 6.4** (top charts), where filtering out hydrodynamic pressure fluctuations had minimal effect. Within this frequency range, the acoustic contribution is more closely matched to the glass bending wavelength than the hydrodynamic contribution, and therefore plays the dominant role in noise transmission.

In contrast, the behavior at low frequencies differs significantly. The bottom charts in **Figure 6.5** offer a zoomed-in view of the low-frequency range (20 Hz–500 Hz) from the full spectra (20 Hz–5000 Hz) shown in the top charts. Within this lower frequency range, the hydrodynamic wavenumber (black line) becomes more closely aligned with the glass bending wavelength than the acoustic wavenumber, for both the side and rear windows. Corresponding hydrodynamic coincidence frequencies occur at approximately 80 Hz for the side window and 110 Hz for the rear window. As a result - opposite to what is observed at mid and high frequencies - hydrodynamic effects are expected to play a more significant role in noise transmission than the acoustic contribution. This is again due to a more efficient interaction with the window's structural eigenmodes near these coincidence frequencies. Supporting this, **Figure 6.4** (bottom charts) shows that filtering out the hydrodynamic pressure component leads to a marked underestimation of interior noise levels - below 360 Hz in the side mirror region and below 500 Hz in the roof spoiler region.

It must be noted, that for the side mirror case, slight deviations from experimental results persisted below 280 Hz for interior noise originating from the side mirror region, even when hydrodynamic contributions were considered (**Figure 6.4**, left). In contrast, excellent accuracy was achieved for interior noise originating from the roof spoiler region when accounting for hydrodynamic contributions (**Figure 6.4**, right).

To understand the origins of the observed deviations in the interior noise predictions, a step-by-step evaluation of the results is necessary, following the sequence of the simulation process. Firstly, the surface microphone results presented in **Figure 5.12** and **Figure 5.13** demonstrated high accuracy in modeling total pressure fluctuations on the windows within the low-frequency range (100 - 500 Hz). This confirmed that the first two steps of the simulation process (CFD and FEM) reliably model this frequency range, ensuring that the input for the third step (VAM) was highly accurate. Deviations were observed only in the third step, specifically in the interior noise predictions for the side mirror case, suggesting that the discrepancies likely arose from the modeling of transmission phenomena. These deviations may be attributed to the free-vibrating boundary condition implemented for the window, which does not fully represent the actual clamping and sealing systems. As shown earlier, window's boundary conditions significantly influence the glass eigenmodes, affecting the transmission behavior and the resulting SPL inside the cabin. However, it is important to note that such deviations are not observed in the roof spoiler case, where the model demonstrated high accuracy even at low frequencies. This indicates that these discrepancies may arise only in specific scenarios, influenced not only by flow conditions but also by the structural characteristics of the window.

Furthermore, significant deviations were also observed below 200 Hz in the validation of acoustic fluctuations propagated in the exterior field (**Figure 5.14** and **Figure 5.15**). Unlike the hydrodynamic-related discrepancies in the interior SPL prediction for the side mirror case (**Figure 6.4**, left), these deviations, involving both the side mirror and roof spoiler cases, likely have a different origin. They may be attributed to the limited size of the source regions, which were optimized for computational efficiency in vehicle development applications. In this low-frequency range, the acoustic wavelengths exceed the maximum dimensions of the source regions implemented in the FE models (**Figure 4.6**), potentially reducing the model's accuracy in capturing these frequencies.

Future research could focus on incorporating realistic sealing and clamping systems to improve the model and deepen understanding of its behavior at low frequencies. Additionally, exploring the influence of simulation volume size at each step of the process could provide valuable insights and further enhance the model's accuracy. However, these refinements lie beyond the scope of this work, which is centered on aeroacoustic analysis within the context of vehicle development applications.

6.1.2 Applications and potential criticalities in vehicle development

The developed simulation process offers the capability to achieve highly accurate results at a relatively low computational cost (details withheld for confidentiality), assuming incompressible flow. The numerical tool has demonstrated robustness and versatility, effectively analyzing different vehicle regions with distinct flow conditions, such as the side mirror and roof spoiler. This flexibility allows targeted investigations of specific components, facilitating local optimizations during the development phase. It supports comparative analyses of various configurations for both exterior and interior noise, and, when combined with wind tunnel results, enables a deeper understanding of interior noise impact, exterior noise propagation, and source characteristics through flow field analysis. Each step of the simulation process delivers distinct and valuable insights: interior noise predictions enable the evaluation of passenger perception; acoustic propagation analysis offers critical information for identifying noise sources and their impact in the surrounding space; and flow field analysis provides insights into noise generation mechanisms while enhancing understanding of aerodynamic behavior. These capabilities make the tool a powerful asset in vehicle aeroacoustic development. However, the efficiency achieved is the result of certain simplifications and assumptions, which introduce limitations and potential criticalities. The tool relies on unsteady CFD simulations (specifically DDES) under the assumption of incompressible flow, justified by the low Mach numbers involved. While this approach reduces computational costs and simplifies the model, it restricts the process to a unidirectional framework. Consequently, the tool can predict exterior noise sources from unsteady flow field data but cannot account for interactions between sound waves and the flow field. Additionally, the simulation model does not account for physical phenomena such as convection, refraction, and diffraction (schematic representation in **Figure 6.6**).

Another potential limitation of the simulation process may arise from the computational cost, which depends on available resources and the specific application. The CFD simulation undoubtedly represents the bottleneck of the process, with the initial step to achieve model convergence incurring the highest CPU cost. Additionally, storing the velocity field in the last phase requires a significant amount of memory (hundreds of GB).

Parallel studies demonstrated that to address these challenges and optimize the overall efficiency, two solutions can be implemented:

- **Initialization of the CFD case in OpenFOAM**: This significantly reduces the physical time required for convergence (from 1.5s to approximately 0.7 s), cutting computational costs by about 30%. The model also demonstrated excellent scalability, with near-linear performance improvements from 500 to 1200 CPUs, indicating potential for further optimization.

- **Use of the idle mode in Actran**: This enables simultaneous execution of CFD and FEM steps. By deleting CFD data already processed by the FEM in real time, memory requirements are reduced by an order of magnitude. Running the final CFD step and the first CAA step in parallel further reduces computational costs by eliminating the need for sequential execution.

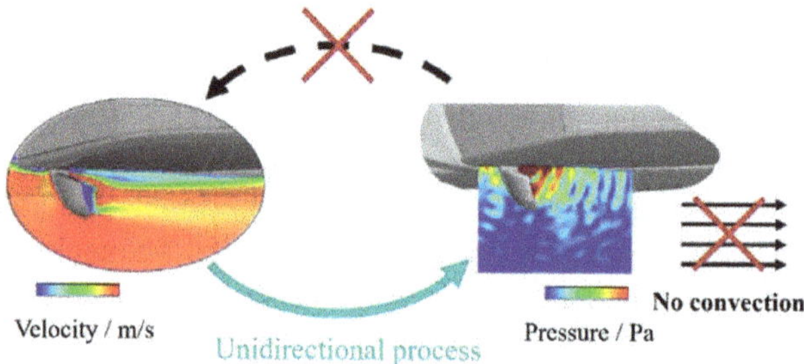

Figure 6.6:   Limitations of the numerical tool: unidirectional process and absence of phenomena like convection, refraction and diffraction

These strategies enhance both computational time efficiency as well as memory efficiency, making the simulation process more practical for resource-intensive applications.

To optimize computational efficiency, the workflow should be tailored to the specific type of analysis. The numerical tool supports both interior and exterior noise investigations. For exterior noise propagation, the APE is required to calculate the sound pressure level, providing critical insights into the exterior field and identifying acoustic sources that influence interior noise. However, as discussed in section 6.1.1, accurate interior noise predictions can be achieved by using the total pressure fluctuation distribution on the window without separating acoustic and hydrodynamic contributions (**Figure 6.4**). This approach reduces computational cost by eliminating the need for the APE step (optional, as detailed in **Figure 4.9**). Accordingly, two distinct workflows are recommended for exterior and interior noise analyses.

Given the hybrid nature of the simulation process, automation is essential to maintain usability. A manual setup would require defining three separate simulation models (CFD, FEM, and VAM) for each configuration, significantly increasing preparation time. Automating the process is highly recommended to reduce overall turnaround time and fully leverage the tool's efficiency.

6.2　Aeroacoustic analysis of the side mirror region

This section provides an in-depth acoustic analysis of the side mirror region, with particular emphasis on its aeroacoustic behavior. Beyond serving as supplementary validation, the analysis also aims to assess the effectiveness of the proposed methodology in addressing key challenges in vehicle development. Specifically, it focuses on the identification of the dominant noise sources and the characterization of the noise generation mechanisms. The investigation focuses on the frequency range of 400 Hz to 5000 Hz, which corresponds to the operational range of the employed numerical tool. Frequencies below 400 Hz were excluded due to known limitations in the tool's accuracy at lower frequencies, as well as their limited relevance in standard vehicle development practices, where such low-frequency contributions are typically deemed negligible for the side mirror region.

6.2.1　Aeroacoustic sources and correlation to the interior noise

A detailed analysis of potential acoustic criticalities in the side mirror region was carried out by investigating the external noise sources in this area. This was achieved through experimental measurements using two beamforming techniques. Standard beamforming was employed to identify and localize the dominant exterior noise sources around the vehicle. In parallel, contribution beamforming was used to correlate these external sources with interior cabin noise, providing dB maps that highlight which sources contribute most significantly to the interior acoustic. In addition to the experimental data, simulation-based analyses were integrated into the investigation. These simulations predictions not only provided further validation of the results but also demonstrated the potential of the numerical approach to be incorporated into the early stages of vehicle development, where physical prototypes may not yet be available, thus offering a viable alternative to experimental testing.

Figure 6.7 presents the results of standard beamforming and simulation predictions using a third-octave band visualization. Beamforming calculations were performed on two intersecting planes (vertical and horizontal) passing through the side mirror to detect sources within this region. To exclude noise from regions outside the simulation model, the analysis was restricted exclusively on the side mirror area. The simulated acoustic field was evaluated on the same horizontal plane as the experimental data.

This analysis aims to provide a qualitative assessment of the primary source locations, while quantitative SPL analysis was covered in the validation section. Two different DFT algorithms, one implemented in the array system and the other in Actran, respectively, were used to postprocess the data. For this reason, the presented results and their associated dB scales are intended solely for identifying the most significant sources and should not be used for direct comparison of absolute values between experiments and simulations. Both visualizations display noise distributions ranging from the maximum detected level to 10 dB below it, with lower levels omitted. For brevity, results are shown for four third-octave bands with central frequencies of 800 Hz, 1000 Hz, 1250 Hz, and 1600 Hz.

The standard beamforming results revealed a consistent spatial distribution of dominant noise sources across the entire analyzed spectrum (400 - 5000 Hz). Notably, a high SPL is detected near the lower part of the connection between the side mirror and the window triangle. Similarly, the numerical model identifies a comparable distribution of noise sources over the same frequency range, aligning closely with the experimental findings. As a result, this specific region of the side mirror was selected for further investigation to better understand the origin of these noise sources and their impact on interior noise.

Furthermore, the contribution of exterior sources to interior noise was assessed by contribution beamforming, using a combination of the wind tunnel array and the reference microphone inside the cabin (see **Figure 5.7**). Calculations were performed on the same vertical and horizontal planes used for standard beamforming. For conciseness, the results, depicted in **Figure 6.8**, are also presented for a reduced frequency range of 800–1600 Hz, consistent with **Figure 6.7**. The dB scales are bounded by the maximum value and 10 dB below it.

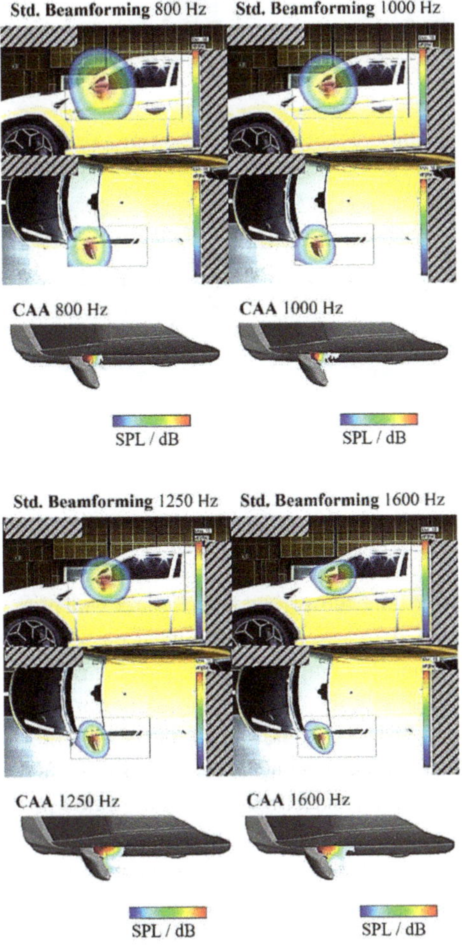

Figure 6.7:   SPL distribution in the side mirror region given for third-octave bands: standard beamforming and CAA results at 800 1000, 1250, 1600 Hz

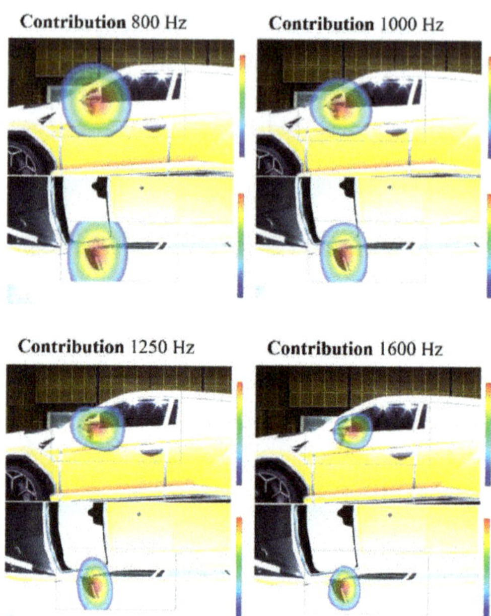

Figure 6.8:   Contribution of the exterior sources to the interior noise at the driver's ear divided by third-octave bands: WT measurements at 800, 1000, 1250 and 1600 Hz

The analysis showed that the aeroacoustic sources identified through beam-forming (see **Figure 6.7**) have a strong impact on the interior noise experienced inside the cabin. By comparing the spatial distribution of exterior noise sources in **Figure 6.7** with the contribution to interior noise shown in **Figure 6.8** a clear correlation emerges: the same regions identified as strong sources outside the vehicle are also responsible for higher noise levels inside. Specifically, the area around the connection between the side mirror and the window triangle was consistently found to be the main contributor to interior noise. This finding holds true across the entire frequency range studied, confirming that this region plays a key role in transmitting aeroacoustic noise into the vehicle cabin.

To further explore the correlation between exterior sources and interior noise, the CAA simulation was employed to map the SPL radiated into the cabin onto the boundary of the VA model. This approach allowed for a clear visualization of the main noise transmission path through the side window. **Figure 6.9** illustrates the SPL distribution on the side window (FEM) and the corresponding SPL radiated into the cabin (VAM), using the third-octave band at 1 kHz as a representative example.

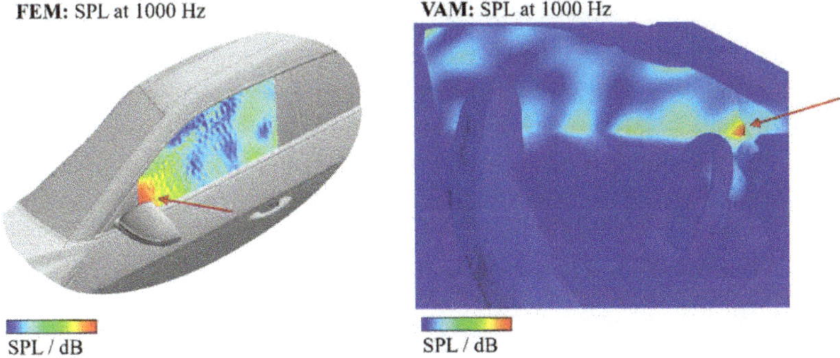

Figure 6.9:　　CAA correlation between exterior and interior noise: SPL distribution on the side window (left) and SPL radiation into the cabin (right) for the third-octave band of 1 kHz

Consistent with the experimental findings, the dominant noise source is located at the base of the triangle, near the connection between the side mirror and the vehicle body. This area shows the highest predicted SPL, as indicated by the red arrows in **Figure 6.9**. It must be noted that the interior door panel and cabin surfaces exhibit very low SPL levels compared to the side window. This occurs because the VA model assumes the glass as the only flexible component and, therefore, the sole medium considered for noise transmission. This assumption aligns with the understanding that windows are the main transmission paths for wind noise originating in the side mirror region, provided the sealing system is functioning effectively.

This comparison provides additional validation of the simulation results and reinforces confidence in the combined numerical-experimental approach, not only by accurately identifying the dominant noise source, but also by guiding effective noise mitigation strategies. For example, design modifications could aim to reduce noise generation at the mirror-body junction or improve damping along the transmission path through the window structure.

6.2.2 Flow field analysis with respect to the acoustic criticalities

The noise source analysis suggests that the main noise generation mechanism in the side mirror region is likely due to vorticity released between the mirror and the vehicle body. Numerical aeroacoustics research [37, 88, 89] has demonstrated that visualizing iso-surfaces of Q, the second invariant of the velocity gradient tensor, is an effective method for identifying regions of dominant vorticity and potential aeroacoustic sources. Accordingly, the main vortex structures in this region were analyzed using the Q-criterion [96], calculated from the transient CFD data extracted during the first phase of the process. **Figure 6.10** illustrates iso-surfaces of averaged Q, colored by vorticity, in the vicinity of the side mirror. Several vortex structures are identified near the side mirror's connection to the vehicle body:

1. Vortex shedding generated by the inner-upper edge of the side mirror.
2. Vorticity released from the lower part of the side mirror housing above its connection to the window triangle.
3. Vorticity released from the lower part of the side mirror housing below its connection to the window triangle.
4. Vorticity generated by the window triangle itself.
5. Vorticity released from the lower part of the side mirror's connection to the window triangle.

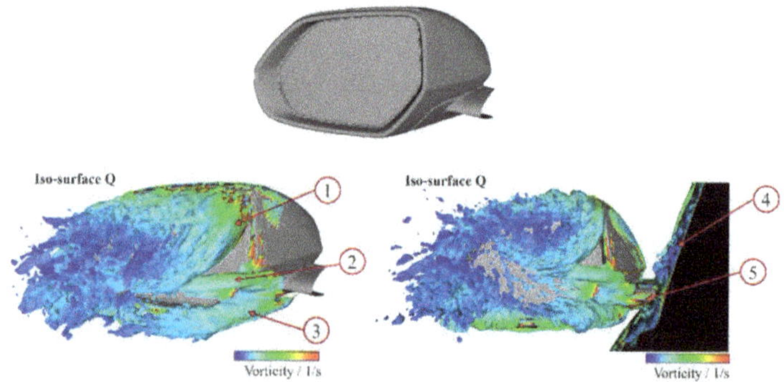

Figure 6.10: Side mirror geometry (top) and Iso-surfaces of time-averaged Q colored by vorticity (bottom, shown from two different viewpoints for clarity)

In aeroacoustic vehicle development, correlations between vortex structures and noise sources [46, 61] are widely used to identify flow regions responsible for significant noise generation mechanisms. In the investigated region, the locations of vortex structures 2, 3 and 5 align closely with the primary noise sources identified by the microphone array and the CAA model (**Figure 6.7**, **Figure 6.8** and **Figure 6.9**). These sources, typically dipoles generated by the interaction of turbulence with solid surfaces (see section 2.2.2), are key contributors to the noise transmitted into the cabin.

In a development scenario focused on reducing interior noise perceived by the driver, it would be crucial to redesign side mirror components in areas where these turbulence structures originate. However, exterior components like side mirrors are the result of interdisciplinary design constraints - including aerodynamics, aeroacoustics, mirror soiling, and vehicle styling. Modifying these components solely to reduce vorticity in the investigated region could compromise other performance aspects. Consequently, such modifications must be evaluated within a broader optimization framework, where trade-offs are quantified and balanced. This underscores the need for integrated design strategies that enable aeroacoustic improvements without compromising the overall functionality and performance of the vehicle.

Overall, the CAA process has demonstrated its suitability for combined analysis with wind tunnel experiments, delivering accurate predictions of the acoustic behavior of specific vehicle regions (e.g., the side mirror) and providing valuable insights into the related flow field.

6.3 Aeroacoustic analysis of the roof spoiler

While extensive research has been conducted on the aeroacoustic behavior of side mirrors, ranging from simplified geometries to real-world vehicle scenarios (see section 3.1.2), there is a notable lack of studies addressing the aeroacoustic phenomena in the rear part of a vehicle, particularly the behavior of roof spoilers. Building on the proven accuracy and applicability of the CAA process, demonstrated through the comprehensive validation and the detailed analysis of the side mirror in the previous section, a similar investigation was conducted for the roof spoiler region.

A comprehensive analysis is presented, including the identification of noise sources, their impact on interior noise, and a flow field analysis aimed at understanding the underlying mechanisms of noise generation. By replicating the approach used for the side mirror, this chapter further validates the versatility of the CAA process for addressing aeroacoustic challenges in different regions of a vehicle.

6.3.1 Aeroacoustic sources and correlation to the interior noise

Compared to the side mirror, the roof spoiler's geometry and position pose greater difficulties in performing acoustic analysis and obtaining accurate experimental and simulation results. Mounted on the car body through lateral connections at the C-pillar, the roof spoiler features a narrow, varying gap between its lower edge and the rear window, with a maximum width of 60 mm. Its rearward location further complicates analysis, as it interacts aerodynamically with nearby elements such as the antenna and the lower spoiler. **Figure 6.11** illustrates an overview of the vehicle's rear, highlighting the components in this region.

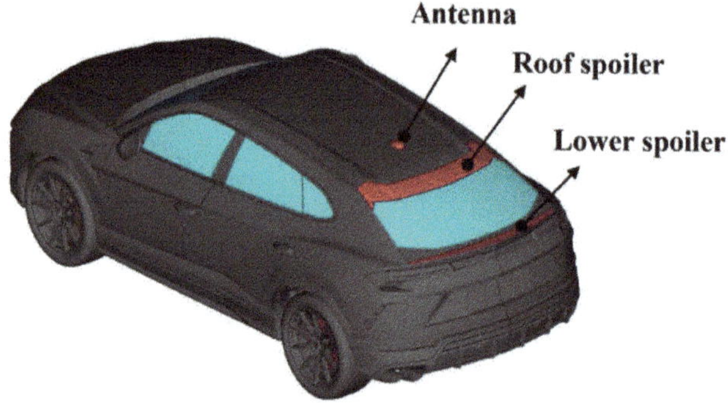

Figure 6.11: Overview of the vehicle's rear aerodynamic components (in red): antenna, roof spoiler, and lower spoiler

To investigate the exterior acoustic field and the propagation phenomena in the roof spoiler region, the beamforming technique was used. Calculations were performed on a horizontal plane located 2 m above the ground, directly above the spoiler. To ensure experimental consistency across different measurement devices, the beamforming plane was positioned at the same Z-coordinate as the exterior microphones used for field validation in section 5.3.2. Results are shown in **Figure 6.13**, focusing on a frequency range of 800 to 1600 Hz in third-octave band visualization as a representative example.

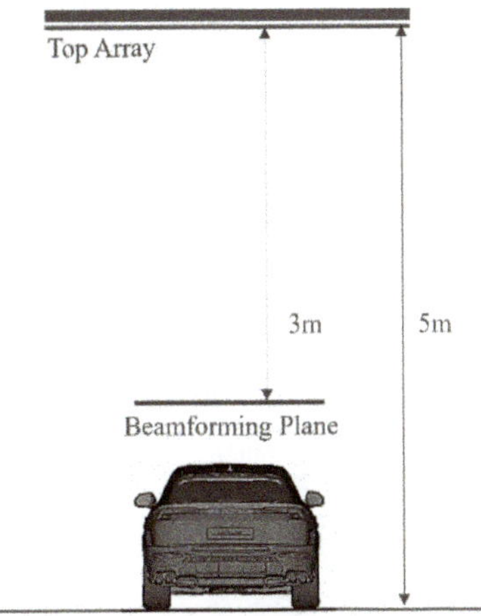

Figure 6.12: Representation of the car located under the wind tunnel array

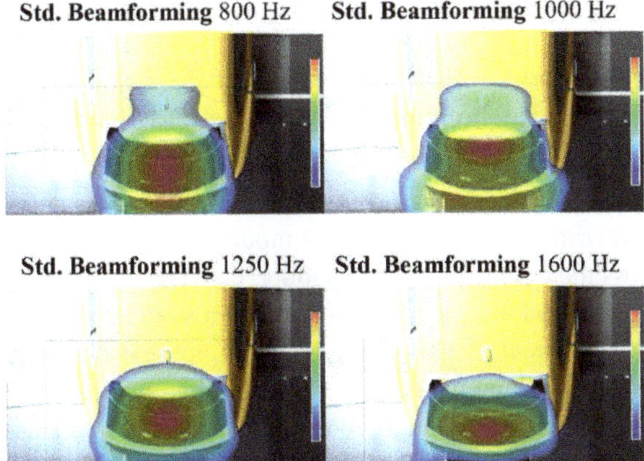

Figure 6.13: SPL distribution in the roof spoiler region divided by third-octave bands: standard beamforming results at 800 1000, 1250, 1600 Hz

The standard beamforming analysis revealed a dominant central noise source across the entire frequency range investigated. To gain a deeper understanding of its propagation into the far field, the CAA process was employed to more precisely analyze the sources generated in the roof spoiler region and their propagation characteristics. The 1 kHz third-octave band was selected as a representative case for this analysis. A qualitative comparison was performed between the standard beamforming results and the numerically predicted SPL distribution on the rear window to identify potential correlations between the exterior SPL distribution and the sources originating within the spoiler gap. This comparison is presented in **Figure 6.14**.

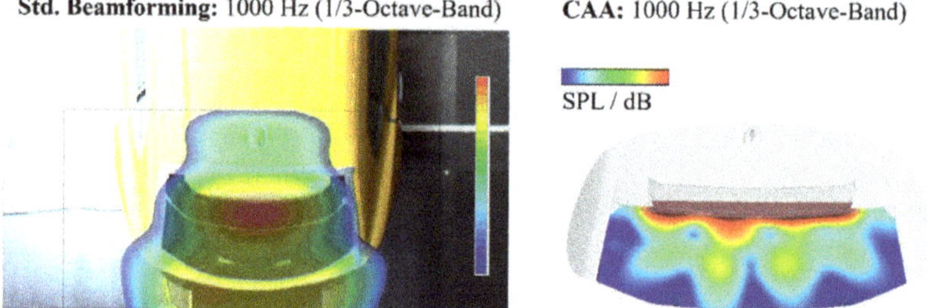

Figure 6.14:   Exterior sound field analysis of the roof spoiler region: WT standard beamforming results (left) vs SPL distribution on the rear window predicted by the CAA process (right)

Unlike the experimental results, the CAA model predicted significant SPL distributed across the entire span of the spoiler, concentrated within the gap between the spoiler and the car body, rather than a single central source as observed in the experiments. This comparison did not reveal any direct correlation between the measurements and simulations. To provide a more detailed analysis and a better understanding of the propagation phenomena, the numerical tool was used to calculate the exterior field on the beamforming plane (see **Figure 6.12**) and a vertical plane positioned directly behind the spoiler, as shown in **Figure 6.15**. The spoiler is depicted as transparent to highlight the sources located behind it, and the exterior virtual microphones used for validation in section 5.3.2 are represented by black square markers.

The analysis revealed significant directivity in the noise propagation behind the spoiler. Three dominant sources were identified: a central source, with noise propagating vertically, and two side sources, where noise diverges laterally during propagation.

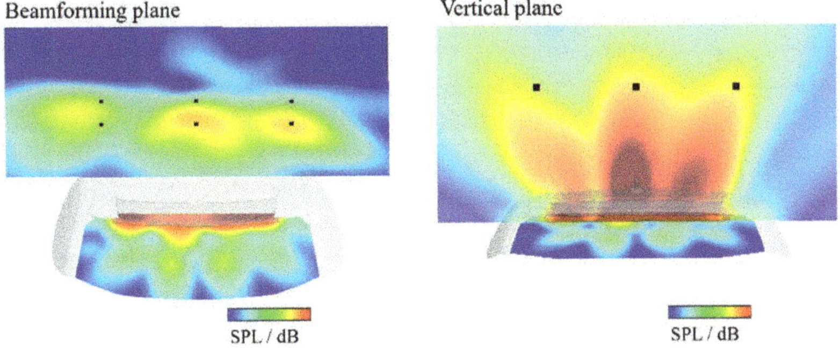

Figure 6.15: CAA exterior field prediction: SPL distribution on the virtual beamforming plane (left) and on a vertical plane located behind the spoiler (right) for a third-octave band of 1 kHz

To evaluate the consistency between the array measurements and the exterior field modeled by the CAA process, the predicted SPL at the top array was examined. **Figure 6.16** presents the SPL distribution (transparent) on the top array, along with the FE model and the virtual microphones in the background used to simulate the exterior field.

The simulation results in the figure above show that only the central noise source was clearly detected by the array, while the lateral sources were only partially identified. This indicates that the lateral sources have strong directivity, meaning they emit sound in specific directions that do not align with the position of the microphone array. As a result, the array was unable to fully capture them. This helps explain the discrepancy observed in **Figure 6.14**, where the experimental data showed a single dominant central source, whereas the simulation predicted multiple sources spread across the span of the spoiler.

In addition, the shape and design of the roof spoiler may also contribute to this issue. The geometry can block or redirect sound waves coming from the gap between the spoiler and the roof, preventing them from reaching the external microphones or the top array (see **Figure 5.4**). This creates an acoustic shadow, which further reduces the identification of these sources. The masking effect caused by the spoiler's geometry is illustrated in **Figure 6.17**.

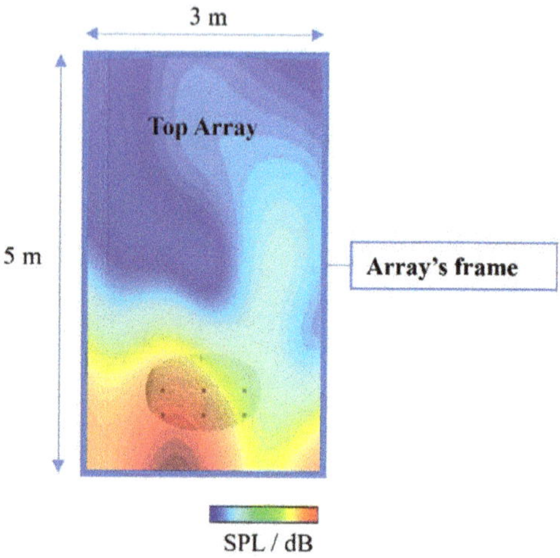

Figure 6.16:   CAA exterior field prediction: SPL distribution on the virtual top array for a third-octave band of 1 kHz

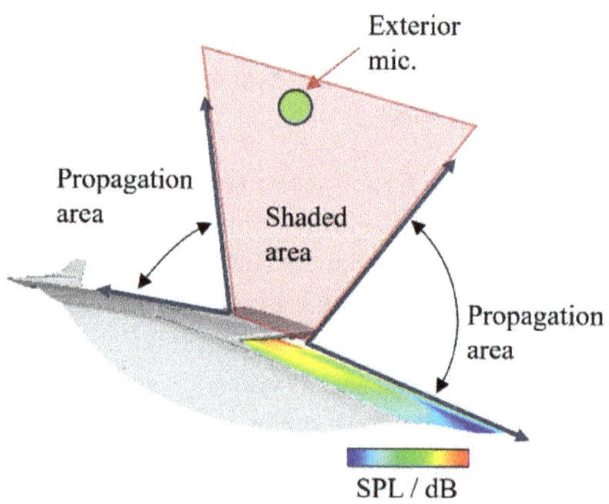

Figure 6.17:   Shaded area due to the masking effect caused by the roof spoiler

In this scenario, where direct comparisons between beamforming measurements and the simulated acoustic field may result in misleading or inconclusive outcomes, the CAA model proves to be a powerful tool for a more detailed investigation of the exterior acoustic field and potential criticalities. Additionally, using exterior microphones placed near the spoiler but outside the turbulent region can provide valuable insights into source propagation, helping to overcome challenges related to source identification. However, while this approach is effective for validation purposes (as detailed in section 5.3.2), it is typically avoided in standard vehicle development due to its time-consuming application.

6.3.2   Aeroacoustic impact of roof spoiler design modifications

The influence of two distinct roof spoiler designs was also examined to evaluate their impact on the exterior generation of noise sources and the resulting effect on passenger noise perception. These included the standard spoiler (**Figure 6.18** left), which has served as the reference throughout the investigations, and a Sport Derivative (SD) configuration (**Figure 6.18** right).

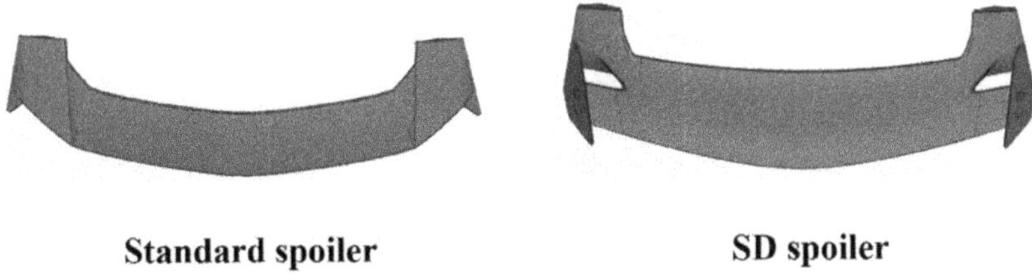

Figure 6.18:   Roof spoiler configurations: Standard (left) and SD (right)

First, the step-by-step validation process outlined in Chapter 5 was applied to the SD spoiler to demonstrate the robustness of each simulation step when subjected to significant geometric changes that may alter both the flow and the acoustics. For brevity, only the predicted SPL at the rear passenger's ear is presented as representative validation in **Figure 6.19**, based on a direct comparison with wind tunnel experiments.

As with the standard spoiler, the investigation of the aeroacoustic impact of the SD spoiler begins with an analysis of the exterior noise sources detected by the microphone array. Standard beamforming was performed on the same beamforming plane used for the standard spoiler, and the results are presented in **Figure 6.20**.

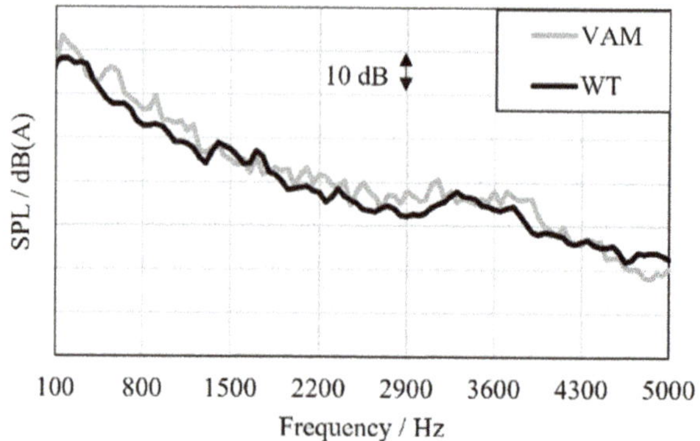

Figure 6.19:   Validation of SPL at and rear passenger's ear induced from the SD spoiler: VAM predictions vs. WT measurements

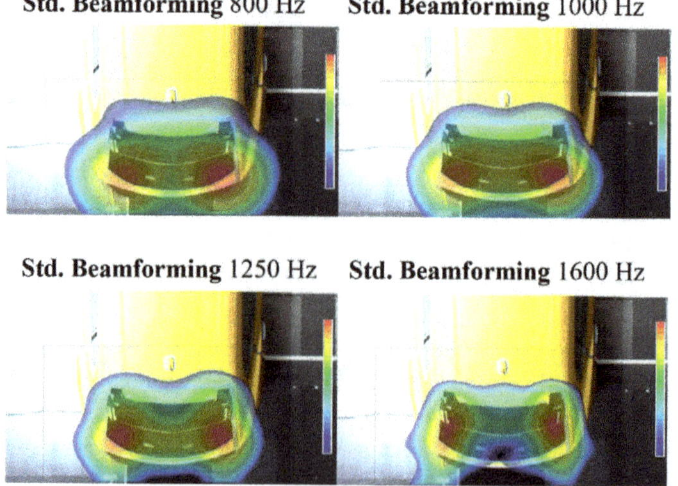

Figure 6.20:   SPL distribution in the SD spoiler region represented by third-octave bands: standard beamforming results at 800, 1000, 1250, 1600 Hz

The implementation of the SD spoiler introduces significant changes to the acoustic field. Unlike the standard spoiler configuration, the dominant sources are now concentrated near the lateral openings of the spoiler design, located at its sides. To better understand the origin of these sources, a qualitative comparison was conducted between the standard beamforming results and the SPL predicted on the rear window. **Figure 6.21** provides an example for the 1 kHz third-octave band.

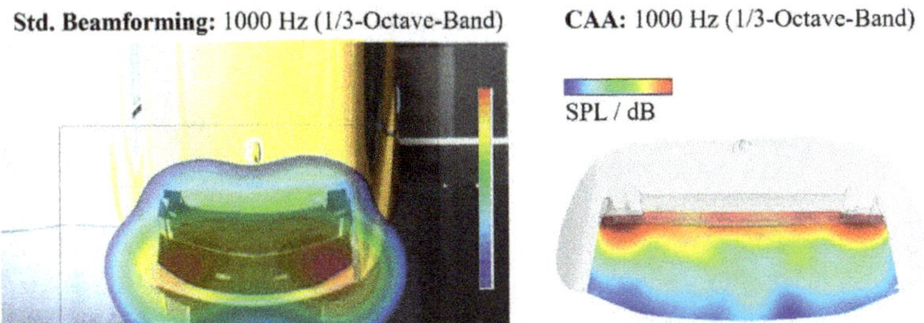

Figure 6.21:   Exterior sound field analysis of the SD spoiler region: WT standard beamforming results (left) vs SPL distribution on the rear window predicted by the CAA process (right)

Unlike the standard spoiler analysis shown in **Figure 6.14**, a direct correlation between the experimental and simulation results was found in this case. The numerical model revealed a similar distribution of critical noise sources located on the sides of the spoiler, consistent with the beamforming results. Additionally, a high SPL is predicted along the entire span of the spoiler, which appears to be detected by the wind tunnel array as well (indicated by the yellow level in **Figure 6.21** left).

This finding suggests that, for the SD spoiler configuration, neither the masking effect nor the directivity of noise propagation significantly impedes the array measurements in this region. Nonetheless, both effects remain relevant considerations in aeroacoustic analysis - particularly near the vehicle periphery -since strong directivity patterns, especially from dipole-type sources, can occur even in the absence of masking. These aspects should be carefully evaluated in future investigations and when applying this tool in vehicle development.

6.3.3   Flow field analysis with respect to the acoustic criticalities

Starting from the prior investigation, the origins of the aeroacoustic sources generated by the two spoiler configurations were analyzed. This analysis used averaged CFD data to develop a comprehensive understanding of the aeroacoustic phenomena and their correlation with the flow field and vortex structures induced by the spoilers. **Figure 6.22** shows the impact of the standard spoiler (top) and SD spoiler (bottom) on the flow field, represented by time-averaged iso-surfaces of Q, colored by vorticity.

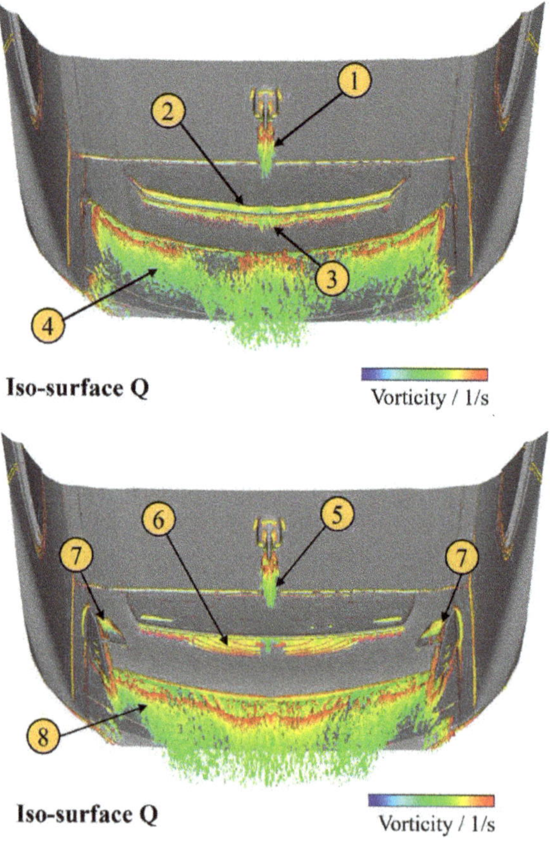

Figure 6.22:   Time averaged flow field induced by the standard spoiler (top) and SD spoiler (bottom): iso-surfaces of Q colored by vorticity

The following vortex structures were identified for the two spoiler configurations:

Standard spoiler:

1. Vorticity generated by the antenna
2. Vorticity originating from the lower part of the leading edge
3. Vorticity originating from the upper part of the leading edge
4. Vorticity caused by separation at the trailing edge, forming the main wake

SD spoiler:

5. Vorticity generated by the antenna
6. Vorticity originating from the upper part of the leading edge
7. Vorticity generated in the lateral openings
8. Vorticity caused by separation at the trailing edge, forming the main wake

The key distinction between the standard and the SD configuration lies in the vorticity induced by the lower part of the leading edge (see vortex structure 2 **Figure 6.22** top). This vorticity is present in the standard spoiler but absent in the SD spoiler due to differences in the leading edge's position and shape. Conversely, the SD configuration introduces lateral openings that generate additional vorticity, released along the side windows through these channels. These differences align closely with the CAA results (compare **Figure 6.14** and **Figure 6.21**), providing further insight into the noise generation mechanisms identified by wind tunnel beamforming and predicted by the simulations.

Figure 6.23 further illustrates these mechanisms by comparing the standard beamforming results (displayed in third-octave bands) with the turbulent kinetic energy (TKE) predicted by CFD on a vertical plane intersecting the spoilers. The vehicle is shown from a top view for the standard beamforming (left) and from a rear view for the CFD analysis (right), highlighting the distinct flow characteristics of each configuration. The same SPL and TKE scales were applied to both models to ensure a consistent basis for comparison.

For the SD spoiler, the high vorticity generated in the lateral channels was identified as the primary noise source. The combined investigations using beamforming, CAA, and CFD consistently highlighted this acoustic phenomenon, directly induced by localized turbulence.

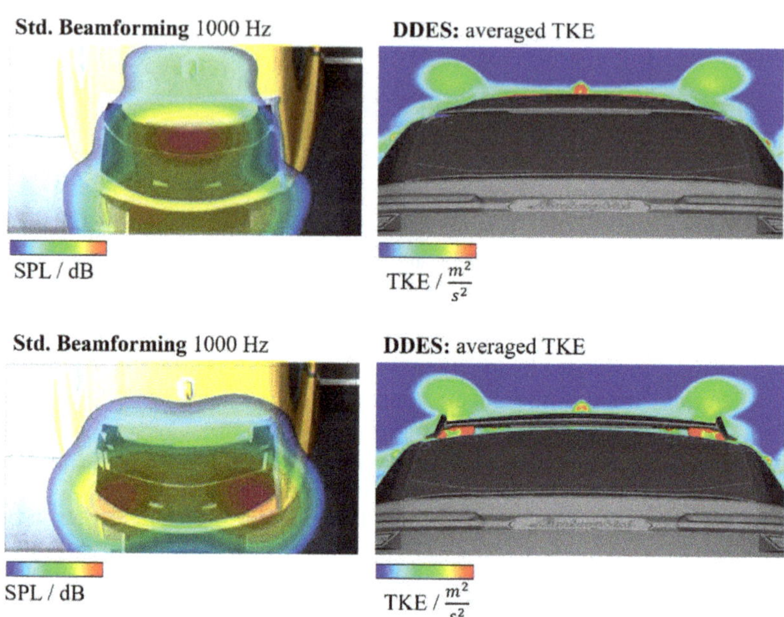

Figure 6.23:   Comparison between the standard beamforming results and the averaged TKE modeled by DDES in the roof spoiler region: standard spoiler (top) and SD spoiler (bottom)

In contrast, the standard spoiler was found to exhibit significantly more complex aeroacoustic behavior, which could only be addressed through a comprehensive analysis. The TKE visualization revealed strong interactions between the antenna's wake and the standard spoiler's leading edge, resulting in high turbulent kinetic energy at the center above the leading edge. This interaction likely generated a prominent dipole, a phenomenon commonly observed in road vehicles. However, by combining beamforming and simulation results - particularly the SPL distribution on the rear window and far-field propagation shown in **Figure 6.15** and **Figure 6.16** - it became clear that the dominant noise generation mechanism was of a different nature. It originated primarily from the turbulence generated at the lower part of the leading edge (turbulence structure 2 in **Figure 6.22**), which then interacted with both the spoiler and the rear glass.

This analysis highlights the value of combining simulation and experimental methods, particularly in cases where a comprehensive understanding of aeroacoustic phenomena is essential for achieving meaningful improvements. While either simulation or wind tunnel testing may be sufficient in many scenarios, relying solely on one method can, in complex cases, result in incomplete or potentially misleading conclusions. The hybrid approach proves especially useful in providing complementary insights into both the flow and acoustic fields, with each contributing essential information. These combined insights are crucial in addressing challenging aeroacoustic issues and supporting effective vehicle development.

7 Conclusions and Outlook

The present thesis research introduces an effective and efficient simulation process for the aeroacoustic analysis of road vehicles, integrating Computational Fluid Dynamics (CFD), Finite Element (FE), and Vibro-Acoustic (VA) models into a hybrid methodology suitable for vehicle development scenarios. Detailed documentation of each simulation model ensures the replicability of the process, even with alternative Navier-Stokes-based CFD or FEA software.

The numerical tool has demonstrated robustness and versatility across a wide frequency range (100 - 5000 Hz), effectively analyzing different vehicle regions with distinct flow conditions, such as the side mirror region and the roof spoiler region. It was proven to facilitate targeted component optimizations during the development phase. Its capability to support comparative analyses of exterior and interior noise configurations, combined with wind tunnel results, enables a deeper understanding of noise propagation, source characteristics, and passenger perception. Each step of the process offers distinct insights: CFD flow field analysis clarifies noise generation mechanisms and improves aerodynamic understanding, FE acoustic propagation analysis identifies noise sources, and the VA model predicts the impact of noise within the vehicle cabin. These capabilities make the tool a powerful asset in vehicle aeroacoustic development.

One of the key advantages of this process is its adaptability to focus on specific frequency ranges. By tailoring the FE computations to the frequency domain of interest, computational costs can be significantly reduced. This is particularly beneficial for low- and mid-frequency investigations, where coarser time and space discretization are sufficient. Additionally, the size of the investigation area can be adjusted based on available computational resources, although including the entire vehicle in a single simulation remains infeasible.

Overall, the process is highly scalable, efficiently utilizing over 1000 cores, and shows promising turnaround times for standard vehicle development scenarios. However, the efficiency achieved through simplifications and assumptions introduces limitations. The reliance on unsteady CFD simulations under incompressible flow assumptions simplifies the model and reduces computational cost. However, it prevents the simulation from accounting for the effects of the acoustic field on the flow field. As a result, physical phenomena resulting from reciprocal interactions - such as acoustic feedback - are not predicted in the model. Computational cost remains a challenge, particularly in the CFD phase, which demands significant CPU resources and memory for storing velocity fields. Furthermore, the hybrid nature of the process, requiring three distinct simulation models (CFD, FEM, and VAM), underscores the importance of automation to reduce preparation time, optimize turnaround, and fully exploit the tool's potential for practical vehicle development scenarios.

Each stage of the simulation process - CFD, FEM, and VA - was individually validated against wind tunnel measurements to ensure consistency and accuracy. The experimental setup was designed to allow direct comparison with simulation results and provides a structured, replicable framework for validating hybrid numerical methods. This framework is not limited to the current application; it can be extended to assess other methodologies based on similar hybrid approaches. Additionally, it can be reused within the same methodology to validate simulations in different regions of the vehicle or for different vehicle models. While this setup is time-intensive and less suited for standard vehicle development, it can be optimized and streamlined for critical measurements to balance validation requirements with the time efficiency typically required by vehicle development phase.

The main noise generation mechanisms at the side mirror and roof spoiler were identified through a combined analysis of simulations and wind tunnel tests. Notably, the investigation of the roof spoiler region, highlighting interactions between components such as the antenna, roof spoiler, and lower spoiler, addresses a gap in the state of the art. Complex phenomena, such as a masking effect caused by the spoiler geometry, were identified, highlighting the challenges of accurately detecting sources using microphones positioned above the vehicle. This study underscores the critical importance of integrating simulations with experiments to achieve a comprehensive understanding of aeroacoustic phenomena. Relying on either method alone risks incomplete or misleading conclusions.

To further enhance the applicability of the developed process, several areas of future research are proposed. First, it would be valuable to validate the simulation process under a broader range of flow conditions, including yaw angles (-20° to 20°) and speeds (90 - 220 kph), reflecting standard wind tunnel testing scenarios. Extending the validation would also provide a more comprehensive assessment of the process's robustness.

Expanding the frequency range of the simulations to include frequencies below 100 Hz could be particularly beneficial in areas where low-frequency contributions significantly affect interior noise, such as the underbody. Furthermore, including the sealing systems into the simulation model would offer insights into overall noise optimization. However, this would increase computational costs, particularly in CFD, as it requires resolving small gaps between windows and pillars.

Another promising avenue for investigation involves conducting distinct simulations around the entire vehicle to model the Pressure Fluctuation Level (PFL) across all windows. This approach would enable a single VA model to incorporate the entire cabin, accounting for all windows simultaneously and providing a more comprehensive analysis. Comparing these results with a fully taped vehicle configuration in the wind tunnel could offer valuable insights into passenger noise perception, yielding a more realistic and holistic understanding of cabin acoustics.

In conclusion, this research has developed an effective hybrid simulation process, a robust validation methodology, and a systematic investigative approach that bridges the gap between experimental and computational aeroacoustics. Its proven effectiveness and adaptability position it as a valuable tool for advancing aeroacoustic studies in vehicle development, with significant potential for future refinements and applications.

References

[1] T. Schütz, Hucho - Aerodynamik des Automobils, Springer Fachmedien Wiesbaden, 2013.

[2] N. Oettle and D. Sims-Williams, "Automotive Aeroacoustics: An Overview," *Proceedings of the Institution of Mechanical Engineers. Part D : Journal of automobile engineering,* 2017.

[3] International Energy Agency, Global EV Outlook 2023: Catching up with Climate Ambitions, OECD, 2023.

[4] C. K. W. Tam, "Computational Aeroacoustics: Issues and Methods," *AIAA Journal,* 1995.

[5] S. Marburg and B. Nolte, Computational Acoustics of Noise Propagation in Fluids - Finite and Boundary Elements Methods, Berlin Heidelberg: Springer, 2008.

[6] E. Manoha, S. Redonnet and S. Caro, "Computational Aeroacoustics," *Encyclopedia of Aerospace Engineering,* 2010.

[7] A. Pietrzyk, D. Beskow, D. Moroianu, M. Cabrol and Y. Detandt, "Passenger Car Side Mirror Exterior Noise Simulation and Validation," in *The 22nd International Congress on Sound and Vibration*, Florence, Italy, 2015.

[8] S. Caro, P. Ploumhans and X. Gallez, "Implementation of Lighthill's acoustic analogy in a finite/infinite elements Framework," in *10th AIAA/CEAS Aeroacoustics Conference*, Manchester, UK, 2004.

[9] S. Caro, Y. Detandt and J. Manera, "Validation of a New Hybrid CAA strategy and Application to the Noise Generated by a Flap in a Simplified HVAC duct," in *15th AIAA/CEAS Aeroacoustics Conference*, Miami, Florida, 2009.

[10] B. Ganty, J. Jacqmot, Z. Zhou and C. Jeong, "Numerical Simulation of Noise Transmission from A-pillar Induced Turbulence into Simplified Car Cabin," *SAE Technical Paper 2015-01-2322,* 2015.

[11] M. J. Lighthill, "On Sound Generated Aerodynamically I. General theory," *Proceedings of the Royal Society,* vol. 211, p. 564–587, 1952.

[12] M. Lighthill, "On sound generated aerodynamically II. Turbulence as a source of sound," *Proceeding of the Royal Society A,* vol. 222, p. 1–32, 1954.

[13] R. Ewert and W. Schröder, "Acoustic perturbation equations based on flowdecomposition via source filtering," *Journal of Computational Physics,* 2003.

[14] C. A. Perugini, G. Arzilli, A. Torluccio, R. Blumrich and A. Wagner, "Application of a newly implemented numerical tool for aeroacoustic vehicle development," *Journal of Tongji University,* 2022.

[15] C. A. Perugini, G. Arzilli, A. Torluccio, R. Blumrich and A. Wagner, "An efficient hybrid computational aeroaocustic process: validation and applications for vehicle development," in *28th AIAA/CEAS Aeroacoustics 2022 Conference*, Southampton, UK, 2022.

[16] C. A. Perugini, U. Riccio, A. Torluccio, R. Mohr, R. Blumrich and A. Wagner, "An Efficient Hybrid Computational Process for Interior Noise Prediction in Aeroacoustic Vehicle Development," *SAE Technical Paper 2023-01-1120, 2023, doi:10.4271/2023-01-1120.*

[17] J. D. Anderson, Fundamentals of aerodynamics (6th ed.), McGraw-Hill Education, 2016.

[18] C. Hirsch, Numerical Computation of Internal and External Flows: Fundamentals of Computational Fluid Dynamics, ELSEVIER, 2007.

[19] S. Glegg and W. Devenport, Aeroacoustics of Low Mach Number Flows, ELSEVIER, 2017.

[20] M. Rieutord, Fluid Dynamics: An Introduction, Springer, 2015.

[21] M. W. Evans and F. H. Harlow, "The Particle-in-Cell Method for Hydrodynamic Calculations," Scientific La- boratory Report, Los Alamos, 1957.

[22] F. Moukalled, L. Mangani and M. Darwish, The Finite Volume Method in Computational Fluid Dynamics: An Advanced Introduction with OpenFOAM® and Matlab, Springer, 2016.

[23] O. Reynolds, "On the Dynamical theory of Incompressible Viscous Fluids and the Determination of the Criterion," *Philosophical Transactions of the Royal Society of London.,* 1895.

[24] D. Wilcox, Turbulence Modeling for CFD, DCW Industries, Inc., 2006.

[25] S. B. Pope, Turbulent Flows, Cornell Univeristy: Cambridge University Press, 2000.

[26] J. Bredberg, "On the Wall Boundary Condition for Turbulence Models," Department of Thermo and Fluid Dynamics CHALMERS UNIVERSITY OF TECHNOLOGY, Göteborg, Sweden, 2000.

[27] H. Tennekes and J. Lumley, A First Course in Turbulence, Cambridge: Massachusetts Institute of Technology,, 1972.

[28] P. R. Spalart, "Detached-Eddy Simulation," *Annual Review of Fluid Mechanics,* 2009.

[29] U. Piomelli, "Wall-layer models for large-eddy simulations," *Progress in Aerospace Sciences,* vol. 44, no. 6, pp. 437-446, 2008.

[30] P. R. Spalart, W.-H. Jou, M. Strelets and S. R. Allmaras, "Comments on the Feasibility of LES for Wings, and on a Hybrid RANS/LES Approach," in *Advances in DNS/LES: Direct numerical simulation and large eddy simulation,* 1997.

[31] P. Sagaut, Large Eddy Simulation for Incompressible Flows: An Introduction, Springer, 2006.

[32] P. Spalart and S. Allmaras, "A one-equation turbulence model for aerodynamic flows," in *AIAA Paper 92-0439*, 1992.

[33] P. R. Spalart, S. Deck, M. L. Shur, K. D. Squires, M. K. Strelets and A. Travin, "A new version of detached-eddy simulation, resistant to ambiguous grid densities," *Theoret. Comput. Fluid Dynamics* , vol. 20, pp. 181-195, 2006.

[34] M. Shur, S. P. R. M. Strelets and A. Travin, "A hybrid RANS-LES approach with delayed-DES and wall-modelled LES capabilities," *International Journal of Heat and Fluid Flow,* 2008.

[35] M. Islam, F. Decker, E. de Villiers, A. Jackson and e. al., "Application of Detached-Eddy Simulation for Automotive Aerodynamics Development," *SAE Technical Paper,* 2009.

[36] P. Ekman, D. Wieser, T. Virdung and M. Karlsson, "Assessment of hybrid RANS-LES methods for accurate automotive aerodynamic simulations," *Journal of Wind Engineering and Industrial Aerodynamics,* vol. 206, 2020.

[37] M. Hartmann, J. Ocker, T. Lemke, A. Mutzke, V. Schwarz, H. Tokuno, R. Toppinga, P. Unterlechner and G. Wickern, "Wind Noise Caused by the Side-Mirror and A-Pillar of a Generic Vehicle Model," in *18th AIAA/CEAS Aeroacoustics Conference (33rd AIAA Aeroacoustics Conference),* 2012.

[38] J. Fröhlich and D. von Terzi, "Hybrid LES/RANS methods for the simulation of turbulent flows," *Progress in Aerospace Sciences,* vol. 44, no. 5, pp. 349-377, 2008.

[39] P. Tucker, "Computation of unsteady turbomachinery flows: Part 2— LES and hybrids," *Progress in Aerospace Sciences,* vol. 47, no. 7, pp. 546-569, 2011.

[40] Y. Tominaga, "CFD simulations of turbulent flow and dispersion in built environment: A perspective review," *Journal of Wind Engineering & Industrial Aerodynamics,* vol. 249, 2024.

[41] Y. Hoarau, S. Peng, D. Schwamborn and A. Revell, "Progress in Hybrid RANS-LES Modelling," in *Papers Contributed to the 6th Symposium on Hybrid RANS-LES Methods*, Strasbourg, France, 2016.

[42] V. Ananthan, N. Ashton, N. Chadwick, M. Lizarraga and D. Maddix, "Machine Learning for Road Vehicle Aerodynamics," in *WCX SAE World Congress Experience*, 2024.

[43] M. Helfer, "General Aspects of Vehicle Aeroacoustics," in *Leactures Series "Road Vehicles Aerodynamics"*, Rhode-St.-Genese, Belgien: Von Karman Institute, 2005.

[44] M. P. Norton and D. G. Karczub, Fundamentals of noise and vibration analysis for engineers, Cambridge University Press, 1989.

[45] A. George, "Automobile Aeroacoustics," in *12th Aeroacoustic Conference, American Institute of Aeronautics and Astronautics*, 1989.

[46] M. Howe, Theory of Vortex Sound, Cambridge: Cambridge University Press, 2002.

[47] M. Howe, "Lectures on the Theory of Vortex-Sound," in *Sound-Flow Interactions*, Springer Link, 2022, pp. 31-111.

[48] J. Schmalz and W. Kowalczyk, "Implementation of Acoustic Analogies in OpenFOAM for Computation of Sound Fields," *Open Journal of Acoustics,* vol. 5, pp. 29-44, 2015.

[49] M. Goldstein, Aeroacoustics, New York: McGraw-Hill International Book Co., 1976.

[50] A. Oberai, F. Roknaldin and T. Hughes, "Computation of Trailing-Edge Noise due to Turbulent Flow over an Airfoil," *AIAA Journal,* vol. 40, pp. 2206-2216, 2002.

[51] A. Oberai, F. Roknaldin and T. Hughes, "Computational procedures for determining structural-acoustic response due to hydrodynamic sources," *Computer Methods in Applied Mechanics and Engineering,* 2000.

[52] Free Field Technologies SA, ACTRAN 2020 User's Guide - Volume 1, 2019.

[53] S. Schoder and M. Kaltenbacher, "Hybrid Aeroacoustic Computations: State of Art and New Achievements," *Journal of Theoretical and Computational Acoustics,* vol. 27, p. 33, 2019).

[54] A. Golota, L. Alimonti and D. Blanchet, "Prediction of interior SPL caused by the Wind Noise," in *DAGA* , Aachen, Germany, 2016 .

[55] H. Frank and C. Munz, "Direct aeroacoustic simulation of acoustic feedback phenomena on a side-view mirror," *Journal of Sound and Vibration,* vol. 371, pp. 132-149, 2016.

[56] L. Mutnuri, S. Senthooran, R. Powell, Z. Sugiyama and D. Freed, "Computational Process for Wind Noise Evaluation of Rear-View Mirror Design in Cars," in *SAE 2014 World Congress & Exhibition,* 2014.

[57] N. Curle, "The influence of solid boundaries upon aerodynamic sound," *Proceedings of the Royal Society of London. Series A, Mathematical and Physical Sciences,* vol. 231, pp. 505-514, 1955.

[58] J. E. a. H. D. L. Ffowcs Williams, "Sound generation by turbulence and surface," *Philosophical Transactions of the Royal Society of London. Series A, Mathematical and Physical Sciences,* vol. 264, pp. 321-342, 1969.

[59] ,. W. Morgan, "The Kirchhoff formula extended to a moving surface," *The Philosophical Magazine: A Journal of Theoretical Experimental and Applied Physics: Series 8,* vol. 9, no. 55, pp. 141-161, 1930.

[60] H. Ribner, "Aerodynamic sound from fluid dilatations; A theory of the sound from jets and other flow," Technical Report, Institute for Aerospace Studies, University of Toronto, 1962.

[61] A. Powell, "Theory of vortex sound," *The Journal of the Acoustical Society of America 36,* pp. 177-195, 1964.

[62] M. Howe, "Contributions to the theory of aerodynamic sound, with application to excess jet noise and the theory of the flute," *J. Fluid Mech.,* vol. 71, p. 625–673, 1975.

[63] W. Möhring, "On vortex sound at low Mach number," *J. Fluid Mech.,* vol. 85, p. 685–691, 1978.

[64] P. E. Doak, "Fluctuating total enthalpy as a generalized acoustic field," *Acoust. Phys.,* vol. 41, p. 677–685, 1995.

[65] J. C. Hardin and D. S. Pope, "An acoustic/viscous splitting technique for computational aeroacoustics," *Theor. Comput. Fluid Dyn.,* vol. 6, p. 323–340, 1994.

[66] W. Z. Shen and J. Sorensen, "Comment on the aeroacoustic formulation of Hardin and Pope," *AIAA Journal,* vol. 37, p. 141–143, 1999.

[67] C. Munz, M. Dumbser and S. Roller, "Linearized acoustic perturbation equations for low Mach number flow with variable density and temperature," *Journal of Computational Physics,* vol. 224, no. 1, pp. 352-364, 2007.

[68] C. Munz, M. Dumbser and M. Zucchini, "The multiple pressure variables method for fluid dynamics and aeroacoustics at low Mach numbers," *Numerical Methods for Hyperbolic and Kinetic Problems,* vol. 7, p. 335–359, 2003.

[69] W. De Roeck and W. Desmet, "Accurate CAA-Simulations using a Low-Mach Aerodynamic/Acoustic Splitting Technique," in *15th AIAA/CEAS Aeroacoustics Conference,* 2009.

[70] D. Blanchet and A. Golota, "Combining Modeling Methods to Accurately Predict Wind Noise Contribution," *SAE Technical Paper 2015-01-2326,* 2015.

[71] J. Kralicek and D. Blanchet, "Windnoise: Coupling Wind Tunnel Test Data or CFD simulation to Full Vehicle Vibro-Acoustic Models," in *DAGA,* Duesseldorf, Germany, 2011.

[72] P. Shorter, V. Cotoni and D. Blanchet, "Modeling interior noise due to fluctuating surface pressures from exterior flow," in *ISMA*, Leuven, Belgium, 2012.

[73] D. Blanchet and A. Golota, "Validation of a wind noise source characterization method for vehicle interior noise prediction," in *ISMA*, 2014.

[74] P. K. Banerjee and R. Butterfield, Boundary element methods in engineering science, New York: McGraw-Hill Book Company, 1981.

[75] F. Fahy, Sound and Structural Vibration, London: Academic Press, 1985.

[76] R. Lyon, Statistical Energy Analysis of Dynamical Systems: Theory and Applications, London: MIT Press, 1976.

[77] B. Lokhande, S. Sovani and J. Xu, "Computational Aeroacoustic Analysis of a Generic Side View Mirror," *SAE Technical Paper 2003-01-1698,* 2003.

[78] R. Hold, A. Brenneis, A. Eberle, V. Schwarz and R. Siegert, "Numerical simulation of aeroacoustic sound generated by generic bodies placed on a plate. I. Prediction of aeroacoustic sources," in *5th AIAA/CEAS Aeroacoustics Conference and Exhibit*, Bellevue,WA,U.S.A., 1999.

[79] R. Siegert, V. Schwarz and J. Reichenberger, "Numerical simulation of aeroacoustic sound generated by generic bodies placed on a plate. II - Prediction of radiated sound pressure," in *5th AIAA/CEAS Aeroacoustics Conference and Exhibit*, Bellevue,WA,U.S.A., 1999.

[80] B. Khalighi, A. Snegirev, J. Shinder, S. Lupuleac and C. K.-B., "Simulations of flow and noise generated by automobile outside rear-view mirrors," *International Journal of Aeroacoustics,* vol. 11, no. 1, 2012.

[81] K. K. Chode, H. Viswanathan and K. Chow, "Noise emitted from a generic side-view mirror with different aspect ratios and inclinations," *Physics of Fluids,* vol. 33, no. 084105, 2021.

[82] X. Chen, S. Wang, Y. Wu, Y. Li and H. Wang, "Experimental and numerical investigations of the aerodynamic noise reduction of automotive side view mirrors," *Journal of Hydrodynamics,* vol. 30, p. 642–650, 2018.

[83] J. Jacqmot, Y. Detandt and G. a. C. Lielens, "Vibro-aero-acoustic simulation of side mirror wind noise and strategies to evaluate pressure contributions," in *ISMA2016*, 2016.

[84] M. Cho, H. Kim, C. Oh, K.-d. Ih and e. al., "Benchmark study of numerical solvers for the prediction of interior noise transmission excited by A-pillar vortex," in *INTERNOISE 2014 - 43rd International Congress on Noise Control Engineering: Improving the World Through Noise Control*, 2014.

[85] F. Mendonca, T. Connelly, S. Bonthu and P. Shorter, "CAE-Based Prediction of Aero-Vibro-Acoustic Interior Noise Transmission for a Simple Test Vehicle," in *SAE Technical Papers 2014-01-0592*, 2014.

[86] M. Oswald and S. Sovani, "Aero-Vibro-Acoustics for Wind Noise Applications," in *DAGA*, Nürnberg, 2015.

[87] S. Lee, S. Lee and C. Cheong, "Development of High-Fidelity Numerical Methodology for Prediction of Vehicle Interior Noise Due to External Flow Disturbances Using LES and Vibroacoustic Techniques," *Apllied Science,* vol. 12, no. 13, 2022.

[88] H. a. Y. Z. Yuan, "Comparison of hydrodynamic and acoustic pressure on automotive front side window," in *Proceedings of the Institution of Mechanical Engineers, Part D: Journal of Automotive Engineering,* 2021.

[89] L. Zhong, Q. Li and Y. a. Y. Wang, "Aerodynamic noise prediction of passenger vehicle with hybrid detached eddy simulation/acoustic perturbation equation method," in *Proceedings of the Institution of Mechanical Engineers, Part D: Journal of Automobile Engineering,* 2018.

[90] S. Chue, "Pressure probes for fluid measurement," *Progress in Aerospace Sciences,* vol. 16, no. 2, pp. 147-223, 1975.

[91] R. a. H. D. Pankhurst, Wind-Tunnel Technique, London: Pitman, 1968.

[92] R. Blumrich, M. Riegel, N. Oettle, H. Vickers and B. Verrecas, "Comparison of Wind Tunnel Array-Based Aeroacoustic Measurement Techniques," in *IVAC18,* 2018.

[93] M. Riegel, R. Blumrich, O. Minck, B. Verrecas, N. Oettle and H. Vickers, "New Large Microphone Array at the FKFS Wind Tunnel," in *12th FKFS-Conference Progress in Vehicle Aerodynamics and Thermal Management,* Stuttgart, Germany, 2019.

[94] C. E. Shannon, "Communication in the Presence of Noise," in *Proceedings of the IRE, 37(1), 10–21,* 1949.

[95] M. Cabrol, Y. Detandt, M. Hartmann and A. Mutzke, "A comparison between the effects of turbulent and acoustic wall pressure fluctuations inside a car," in *18th AIAA/CEAS aeroacoustics conference,* Colorado Springs, 2012.

[96] J. Hunt, A. Wray and P. Moin, "Eddies, Streams, and Convergence Zones in Turbulent Flows," in *Center for Turbulence Research Report,* 1988.

[97] M. Rieutord, Fluid Dynamics: An Introduction, Springer, 2015.

GPSR Compliance
The European Union's (EU) General Product Safety Regulation (GPSR) is a set
of rules that requires consumer products to be safe and our obligations to
ensure this.

If you have any concerns about our products, you can contact us on

ProductSafety@springernature.com

In case Publisher is established outside the EU, the EU authorized
representative is:

Springer Nature Customer Service Center GmbH
Europaplatz 3
69115 Heidelberg, Germany